Optimization Of Traffic Flow Networks

Vom Fachbereich Mathematik

der Technischen Universität Kaiserslautern

zur Verleihung des akademischen Grades

Doktor der Naturwissenschaften

(Doctor rerum naturalium, Dr. rer. nat.)

genehmigte

Dissertation

von

Dipl.–Math. Claus Kirchner

aus Dillhausen

Datum des Einreichens: 12.04.2007
Tag der mündlichen Prüfung: 13.07.2007

Bibliografische Information der Deutschen Nationalbibliothek

Die Deutsche Nationalbibliothek verzeichnet diese Publikation in der Deutschen Nationalbibliografie; detaillierte bibliografische Daten sind im Internet über http://dnb.d-nb.de abrufbar.

ISBN 978-3-8325-1678-9

Logos Verlag Berlin
Comeniushof, Gubener Str. 47,
10243 Berlin
Tel.: +49 030 42 85 10 90
Fax: +49 030 42 85 10 92
INTERNET: http://www.logos-verlag.de

Acknowledgements

I would like to express my deep gratitude to Prof. Dr. Axel Klar for introducing me to the interesting topic of traffic flow phenomena. He supported this work with his advice and encouragement during the last years.

I am particularly indebted to Jun.Prof. Dr. Michael Herty for his time, helpful discussions and invaluable advice that contributed to major parts of this work.

Furthermore, I would like to thank Prof. Dr. Randall LeVeque and his post–docs and graduate students for their support during my research stay at the University of Washington. I was able to get some insights into numerical linear algebra. Additionally, I learned about the latest high–resolution algorithms for conservation laws and could work with the software CLAWPACK and, more specifically, TSUNAMICLAW.

Additionally, I would like to thank Prof. Dr. Reinhard Illner for his hospitality during my two–week stay at the University of Victoria. His ideas and support were priceless for the parts of this work regarding kinetic models.

Subsection 2.2.1 owes its existence and notation to an autumn school I attended at the TU Hamburg in September 2005. The material in the particular subsection was presented by Prof. Dr. Stefan Ulbrich.

Prof. Dr. Sven Krumke contributed major parts on the idea for a preprocessing routine found in appendix B.

I thank Prof. Dr. Lorenzo Pareschi for favorably answering the call to be co–referee.

Finally, I extend my appreciation to all my colleagues of the AG Technomathematik at the University of Kaiserslautern and the Graduiertenkolleg "Mathematik und Praxis" at the University of Kaiserslautern. They provide a stimulating and fruitful working atmosphere.

The work on supply networks (sections 3.1.1 and 3.2.2) was performed in cooperation with S. Göttlich, a fellow PhD student. The parts of the work concerning the modeling, the discrete supply network model and the discrete optimization are her work whereas the adjoint optimization technique and related questions were contributed by me.

Last but not least I would like to thank my family and Liz for their constant encouragement during the past three years.

This work was financially supported by the German Research Foundation (DFG) via a fellowship in the Graduiertenkolleg "Mathematik und Praxis" at the TU Kaiserslautern. Additional funding was provided by the German Academic Exchange Service (DAAD), grants D/05/43739, PPP Kanada and PPP Vigoni.

Kaiserslautern, August 7, 2007   Claus Kirchner

Contents

Chapter 1

Introduction

In the modern world, network flow problems arise in many different situations. For example, many people use a computer network every day for checking email or browsing the web. Production processes in factories involve supply chains, a particular form of a network. Furthermore, many people go to work by bus, train, or car every day and thus use traffic networks. Therefore, networks are important and their study is valuable.

The steady growth of traffic volume, the occurence of heavy traffic jams, and the increasing need of mobility in our society prompt us to consider traffic problems. Traffic flow models of highways differ in certain aspects, in particular in their complexity.

Over the past 70 years lots of ideas have been stated and a variety of approaches have been suggested [85, 95, 22, 91, 46, 73, 72, 7] in order to model traffic flow on a single road. Models of traffic flow are classified according to the level they describe. We distinguish a microscopic, a mesoscopic and a macroscopic scale.

In the microscopic approach each individual car is modeled as well as the interaction among the cars. This idea leads to Follow–the–Leader type models [22, 71, 96] and mathematically results in a (usually large) system of ordinary differential equations. Kinetic or mesoscopic descriptions of traffic flow [74, 66, 50] use distribution functions which model the average behavior of drivers. They usually rely on Boltzmann or Vlasov–Fokker–Planck type equations. Finally, the macroscopic description [85, 91, 48, 7, 51, 81, 82] employs a fluid–dynamic description of traffic flow. This is the approach we will most follow in this thesis.

The first macroscopic model was thought of by Lighthill and Whitham [85] and Richards [95] (LWR-model). Various modifications and extensions to this basic model have been studied and discussed, cf. [91, 6, 40, 27, 7]. Today, fluid dynamic models for traffic flow are appropriate to describe traffic phenomena such as congestion and stop-and-go waves [66, 41, 74]. Recently, there has been intensive research into traffic flow for road networks in the mathematical [19, 34, 62, 53, 35, 54] as well as in the engineering community [81, 82].

The second chapter of this thesis concerns traffic flow networks. In the first section we will give a brief introduction into select models, which found the basis for the following work. Presumably, the first model created to describe traffic flow on a network is discussed in [62]. A more suitable approach was developed by Coclite/Garavello/Piccoli [19]. We will review their model for traffic flow networks in section 2.1.1. It serves as a foundation for the work presented in this thesis. We will also discuss a "second–order" traffic flow model suggested by Aw and Rascle [7].

Once we are familiar with these preliminaries, we will then introduce a promising new approach for modeling traffic flow networks, as suggested in [34], which extends the network model described in [19]. We will use multicommodity flows to incorporate drivers' preferences into a traffic network.

A variety of questions arises regarding this modified model. A major one is the solvability of the additional equations. This matter is addressed in subsection 2.1.3 and more thoroughly in appendix A. We show that we can always construct a weak solution to the system of hyperbolic equations describing the dynamcis on the network. Additionally, the coupling conditions at a junction need to be modified to obtain a correct model. For the case of two classes, a more theoretical analysis is conducted in subsection 2.1.4 that highlights how one can obtain the correct solution at a junction.

Once one has a suitable model for traffic flow one wants the model to influence the behavior of drivers. For example, one objective is to minimize the number of traffic jams in a network in a given time–frame. Among the first papers on traffic network optimization we mention the work [51].

Since the dynamics on a traffic network are governed by a system of hyperbolic equations, we need to deal with optimization problems on networks in which the constraints are given by partial differential equations (PDEs). For certain types of these kind of optimization problems solution methods are known and discussed in [78, 55, 56, 57, 58, 59, 103, 104, 102, 92, 16, 79].

In section 2.2 of this work we formally extend these to a network setting. We present a general framework to compute first–order necessary optimality conditions in subsection 2.2.1. These results are successfully applied to traffic flow and supply networks in subsections 2.2.2 and 3.1.1, respectively.

The proposed procedure is expensive in terms of memory and computational time for large networks. In other types of problems similar difficulties arise and were successfully tackled by an instantaneous control approach [87, 55, 57, 49]. This method which, in general, solely computes suboptimal controls can be successfully applied to traffic networks as the results from subsections 2.2.2 and 2.3.2 indicate.

For the optimization process we use gradient–based methods. However, more efficient algorithms incorporate second order information. For an instantaneous control problem we show in appendix C how the framework introduced in [102] can be used to compute the Hessian for a class of objective functions.

Section 2.3 contains numerical results on the topics treated in section 2.1 and 2.2. Subsection 2.3.1 contains numerical simulations for the new multicommodity model. In particular, we show that the model behaves as intended. An evaluation of the instantaneous control strategy can be found in subsection 2.3.2.

In subsection 3.2.2 we report results on optimal control for supply networks. In particular, we find computational evidence that the formulas we obtain from our general framework from subsection 2.2.1 and subsection 3.1.1 are correct.

Chapter 4 is devoted to kinetic models of traffic flow. In section 4.1 we include an important result concerning the relationship of traffic flow models of different scales. We show that the celebrated macroscopic Aw–Rascle model [7] can be derived from a kinetic, Fokker–Planck type equation. The derivation implies that kinetic traffic flow models might be able to describe traffic phenomena in greater detail than their macroscopic counterparts. Therefore, it is interesting to study kinetic equations for traffic flow in their own right. Consequently, in section 4.2 we conduct a stability analysis for a particular kinetic traffic flow model.

Our result from section 4.1 ties mesoscopic and macroscopic models together. It blends in with earlier approaches to derive macroscopic traffic flow models from kinetic equations [74]. We note that similar results hold for supply chains [4]. A general deduction technique for macroscopic equations from kinetic ones can be found in [84].

Finally, we summarize the main results and give an outlook to open problems and interesting areas for future research.

Chapter 2

Traffic Flow Networks

This chapter addresses traffic flow networks. In section 2.1 we present different models for traffic flow networks. In particular, a reference network model is reviewed. We also show how this model can be extended to incorporate drivers' source–destination preferences.

Ultimately, one wants to minimize traffic jams in a traffic network. Once one has determined a model for simulation one can study certain kinds of optimization problems. This is done in section 2.2. We present a general optimization framework for networks and apply it to the reference network model in the sequel.

We end this chapter in section 2.3 with numerical results concerning the extended reference network model and a particular optimization technique.

2.1 Traffic Flow Modeling

This section is devoted to a vital aspect in applied mathematics: reasonably modeling a real–world process. In this chapter we are concerned with traffic flow. Among the many models for vehicular traffic flow on a single road we are interested in macroscopic models as introduced in [85]. A disscussion and alternatives to macroscopic models can be found in [7, 25, 72, 62, 91, 47] and the references therein.

The extension of macroscopic traffic models to networks yields coupled systems of hyperbolic equations: on each edge of the network we assume a conservation law holds with boundary conditions determined by conditions at the vertices of the network. We will review a network model for vehicular traffic originally proposed by Coclite/Piccoli in subsection 2.1.1. It was one of the first models describing traffic flow on a network.

Naturally, certain effects were not included; in particular, driver preferences are not taken into account. To overcome this shortcoming we are interested in multi–class traffic flow models, i.e., models containing two or more car species, e.g. cars having different destinations, cf. [34]. Such models have been proposed for example in [8, 10, 108, 107, 105] and in [34]. The latter publication introduces a source–destination model based on the LWR equation for a road network and analyzes its mathematical properties [34].

We follow this approach in subsection 2.1.2. The key idea is to separate drivers according to their source destination relations. This leads to additional equations for the so–called commodities. Their solution structure is discussed in subsection 2.1.3 and more thoroughly in appendix A. Additionally, subsection 2.1.2 contains a modification of the coupling conditions at junctions of the reference model. This is necessary to obtain a correct multi–class model.

Subsection 2.1.4 highlights results for the particular case of two classes in a network. We study in detail the solution to Riemann problems at the junctions. This analysis is the key to solving the two–class model (or any other traffic flow model) on a network.

2.1.1 Review of a reference network model

We begin this section with a description of the dynamics on a single road. We review two possible models, the classical LWR–equation [85] and the more sophisticated Aw–Rascle system [7]. Then we will discuss a model devised by Garavello/Coclite/Piccoli that uses the LWR–equation to define traffic flow on road networks. By now there exists also a network formulation based on the Aw–Rascle equations [35, 53, 54].

Traffic Models on a unidirectional road

We briefly summarize the main developments and ideas of macroscopic traffic flow models, i.e., mathematical descriptions of vehicular traffic based on conservation laws. Macroscopic models are formulated in terms of averaged quantities (e.g., the density of cars, i.e., the number of cars over time) rather than dealing with each car individually. The latter idea is usually referred to as microscopic approach.

The basic principle macroscopic models rely on was given by Lighthill and Whitham in 1955, see [85]. They argue as follows: Consider a road with one lane and unidirectional traffic. Let $\rho = \rho(x,t)$ denote the density of cars at a point $x \in \mathbb{R}$ and a time $t \in \mathbb{R}^+$. Each road is modeled as an (unbounded) interval in $\mathbb{R}$. The conservation of the number of vehicles gives rise to the following hyperbolic equation:

$$\partial_t \rho(x,t) + \partial_x(v(x,t)\rho(x,t)) = 0$$

$v(x,t)$ denotes the velocity of cars at the point $(x,t) \in \mathbb{R} \times \mathbb{R}^+$. We need a functional correspondence between v and ρ in order to obtain a closed set of equations. Lighthill and Whitham proposed to assume a local equilibrium distribution $u^e(\rho)$. This distribution u^e is called the *fundamental diagram* and the flux f is defined as:

$$v(x,t) := u^e(\rho(x,t)) \qquad f(\rho) := \rho u^e(\rho) = \rho v(x,t) \qquad (2.1)$$

This implies that all drivers choose their current velocity according to the local densities only. Besides this crucial assumption the obtained equation (2.2) describes many features of traffic flow. In particular, backwards propagating shock waves appear if $\partial_\rho(\rho u^e(\rho)) < 0$. for some ρ. In the following we will skip the argument (x,t) for brevity whenever the intention is clear:

$$\partial_t \rho + \partial_x(\rho u^e(\rho)) = 0 \qquad (2.2)$$

Examples for the choice of u^e are

$$u^e(\rho) \;=\; v_0 \left(\frac{1}{\rho} - \frac{1}{\rho_{\max}} \right)$$

$$u^e(\rho) \;=\; v_{\max} \left(1 - \left(\frac{\rho}{\rho_{\max}} \right)^n \right)$$

For a comparision of these and other models and support through experimental data see [86].

The assumption of a functional correspondence of the form (2.1) is subject to discussion. Measurements indicate that a relation of the form (2.1) only holds for sufficiently small values of the density ρ and that the fundamental diagram is multi–valued in a so called *congested region* [71, 70]. In particular, one can then not define the fundamental diagram as a function of ρ. R. Colombo developed a model that incoporates the idea of a congested and a free–flow region and discusses transitions between the two regions [20, 21]. It is related to the Aw–Rascle–model stated below.

Another natural approach to incorporate multi–valued fundamental diagrams is to determine the mean speed v in (2.1) by an additional partial differential equation which is in general coupled to the conservation law for the densisty ρ. This leads to new models, see [91, 7] for example. Reasonable extensions of this form need to satisfy certain conditions, cf. [25]. E.g., the model proposed in [91] violates a fundamental principle stated in [25]. A car is anisotropic: it responds well to frontal stimuli but does not respond well or similarly to stimuli from behind. Therefore, cars are different from fluid particles and reasonable models need to take this into account.

For future reference we review the Aw–Rascle (AR) model for traffic flow [7] which resolved the problems posed in [25]. Other authors used the Aw–Rascle model as basis for their research [40] and by now the AR model is well–established; recently it was even extended to networks by a series of publications [54, 35, 53].

The original Aw–Rascle model [7] without diffusive effects or relaxation terms is given by the 2×2 system of coupled nonlinear partial differential equations

$$\partial_t \rho + \partial_x (\rho v) \;=\; 0 \tag{2.3a}$$
$$\partial_t (v + p(\rho)) + v \partial_x (v + p(\rho)) \;=\; 0 \tag{2.3b}$$

where the "pressure" $p = p(\rho)$ is a smooth increasing function. In [7] the authors suggest a prototype of the form

$$p(\rho) = \rho^\gamma, \quad \gamma > 0$$

More generally, they require the pressure p to satisfy

$$p(\rho) \sim \rho^\gamma \text{ near } \rho = 0, \gamma > 0 \quad \text{and} \quad \forall \rho : \rho p''(\rho) + 2p'(\rho) > 0$$

With these assumptions they show that the second order system (2.3) is hyperbolic and resolves all modeling problems outlined in [25].

The previous macroscopic road models are useful to describe traffic flow behavior on a single–lane road. Macroscopic and kinetic multi–lane models are discussed in

13

[26, 27, 28, 66], for example. In multi–lane models one naturally needs a lane–change or passing probablility and for each lane an equation which describes the dynamics in this lane. Therefore, these models are more involved and have additional features not present in single–lane road models. To the best of my knowledge, so far there have not been any attempts to consider multi–lane models in a network setup. We will not be concerned with multi–lane models in the sequel.

Models for traffic networks

We briefly recall the traffic flow model for a road network with coupling conditions at the junction as in [19, 49] and refer for details to these references. We consider as a road network a finite, directed graph $G = (V, A)$, where each arc $j \in A$ models a uni-directional road and each vertex $v \in V$ a junction. Furthermore, each road j is modeled by an interval $[a_j, b_j]$ where we allow a_j or b_j to be infinity. The sets δ_v^- and δ_v^+ denote the ingoing and outgoing arcs at a vertex $v \in V$, respectively. The functions $\rho_j(x, t)$ for $j \in A$ describe the density of vehicular traffic on road j. We assume that dynamics of traffic flow follow the Lighthill-Whitham-Richardson equation [85] given by

$$\partial_t \rho_j + \partial_x f_j(\rho_j) = 0 \qquad \forall (x, t) \in [a_j, b_j] \times [0, T] \tag{2.4a}$$

$$\rho_j(x, 0) = \rho_{j,0}(x) \qquad \forall x \in [a_j, b_j] \tag{2.4b}$$

Herein, f_j is the flux function and is assumed to be a differentiable, concave function of ρ. For a discussion of the validity of this model and its properties we refer to [85, 62, 72].

One way to solve systems of hyperbolic equations involves so–called Riemann problems [68, 37, 63, 80, 83, 24]; we will discuss particular instances below.

Definition 2.1.1. *A Riemann problem for a hyperbolic equation consists of the equation under consideration and Cauchy–data of Heaviside–type with $u_l \neq u_r$:*

$$\partial_t u + \partial_x f(u) = 0 \tag{2.5a}$$

$$u(x, 0) = \begin{cases} u_l & x < x_0 \\ u_r & x > x_0 \end{cases} \tag{2.5b}$$

As in [19, 62, 49] we assume that f is concave and has a unique maximum σ_j, cf. figure 2.1:

$$f_j(0) = f_j(\rho_j^{\max}) = 0 \tag{2.6a}$$

$$\exists \sigma_j \in (0, \rho_j^{\max}) \text{ such that } f_j'(\sigma_j) = 0 \wedge (\rho_j - \sigma_j) f_j'(\rho_j) < 0 \quad \forall \rho_j \neq \sigma_j \tag{2.6b}$$

For arcs $j \in A$ that are ingoing or outgoing to the network (i.e., $j = (v_j, w_j) \in A$ with $\delta_{v_j}^- = \emptyset$ or $j = (v_j, w_j) \in A$ with $\delta_{w_j}^+ = \emptyset$) we assume boundary conditions in the sense of [9]. The natural coupling condition resembles Kirchhoff's law and reads

$$\sum_{j \in \delta_v^-} f_j(\rho_j(b_j, t)) = \sum_{j \in \delta_v^+} f_j(\rho_j(a_j, t)), \quad \forall t \geq 0 \tag{2.7}$$

at a junction $v \in V$ with incoming roads $j \in \delta_v^-$ and outgoing roads $j \in \delta_v^+$, respectively.

In order to derive a well–defined problem on the whole network we cite a formulation of a well–defined problem at each *junction*.

Definition 2.1.2. (Weak solutions of a Riemann problem at a junction [62]) *By the definition of a weak solution of the Riemann problem for a junction $v \in V$ we mean a weak solution of the initial value problem (2.4) for the network consisting of the single junction v with $j = 1, \ldots, n$ ingoing and $j = n+1, \ldots, n+m$ outgoing roads, all extending to infinity. The initial data are given by*

$$\rho_{j,0}(x) \quad = \quad \rho_{j,0} \qquad \forall x \in [a_j, b_j], j = 1, \ldots, n+m \tag{2.8}$$

where $\rho_{j,0} = \tilde{\rho}_j \in \mathbb{R}$ are constant.

Assuming a solution in the sense of definition 2.1.2, we can denote this solution at the junction by

$$\bar{\rho}(t) := (\bar{\rho}_1, \ldots, \bar{\rho}_{n+m})(t) := \left(\begin{array}{cccc} \rho_j(x = b_j, t) & j & = & 1, \ldots, n \\ \rho_j(x = a_j, t) & j & = & n+1, \ldots, n+m \end{array} \right)$$

If we pose certain restrictions on the Riemann data, it turns out that $\bar{\rho}(t)$ is independent of time

$$\bar{\rho} = \bar{\rho}(t)$$

and the Riemann problems (2.11) and (2.12) are well–posed, see below.

A weak solution of the network problem is defined in [62]. We want to briefly restate the definition. We introduce test functions ϕ_j on all arcs $j \in A$ which are smooth at the intersections and have compact support in $[a_j, b_j]$. Smoothness across a junctions means, for example, in the case of an ingoing road l and outgoing road k

$$\phi_l(b_l) = \phi_k(a_k) \quad \text{and} \quad \partial_x \phi_l(b_l) = \partial_x \phi_k(a_k) \tag{2.9}$$

Definition 2.1.3. *Consider a network $G = (V, A)$. Then $\{\rho_j\}_{j=1,\ldots,|A|}$ is called a weak solution, if it satisfies*

$$\sum_{j=1}^{|A|} \left(\int_0^\infty \int_{a_j}^{b_j} [\rho_j \partial_t \phi_j + f(\rho_j) \partial_x \phi_j] \, dx \, dt - \int_{a_j}^{b_j} \rho_{j,0} \phi_j(x, 0) \, dx \right) \quad = \quad 0 \tag{2.10}$$

for all families of smooth test functions ϕ_j.

We can obtain a weak solution $\rho_j(x, t) \, j = 1, \ldots, n+m, \, \forall x \in [a_j, b_j], \, \forall t \in \mathbb{R}^+$ at the junction $v \in V$ in the sense of definition 2.1.2 by solving the following Riemann problems:

$$j = 1, \ldots, n$$

$$\partial_t \rho_j + \partial_x f(\rho_j) = 0 \tag{2.11a}$$

$$\rho_{j,0} := \rho_j(x, 0) = \left\{ \begin{array}{ccc} \tilde{\rho}_j & x & < & b_j \\ \bar{\rho}_j & x & = & b_j \end{array} \right\} \tag{2.11b}$$

$$j = n+1, \ldots, n+m$$

$$\partial_t \rho_j + \partial_x f(\rho_j) = 0 \tag{2.12a}$$

$$\rho_{j,0} := \rho_j(x, 0) = \left\{ \begin{array}{ccc} \tilde{\rho}_j & x & > & a_j \\ \bar{\rho}_j & x & = & a_j \end{array} \right\} \tag{2.12b}$$

15

The main question in defining traffic flow on a network is the definition of suitable additional conditions at the junctions to obtain unique values $(\bar{\rho}_1, \ldots, \bar{\rho}_{n+m})$. A suitable procedure was first outlined in [62]. A slightly modified approach is presented in [19]. We will review the latter below as the suggested additional conditions at the junctions form the core of the reference model, cf. [19].

Again, we consider a junction with $j = 1, \ldots, n$ ingoing and $j = n+1, \ldots, n+m$ outgoing roads. In [19] the possible values of $\bar{\rho}_j$, $j \in A$ are restricted in order to obtain the correct wave speeds and thus have $\bar{\rho}_j$ constant in time:

$$
\left.
\begin{array}{llllll}
\bar{\rho}_j \in [\sigma, 1] & \tilde{\rho}_j & \geq & \sigma & j & = 1, \ldots, n \\
\bar{\rho}_j \in \{\tilde{\rho}_j\} \cup [\tau(\tilde{\rho}_j), 1] & \tilde{\rho}_j & \leq & \sigma & j & = 1, \ldots, n \\
\bar{\rho}_j \in [0, \sigma] & \tilde{\rho}_j & \leq & \sigma & j & = n+1, \ldots, n+m \\
\bar{\rho}_j \in [0, \tau(\tilde{\rho}_j)] \cup \{\tilde{\rho}_j\} & \tilde{\rho}_j & \geq & \sigma & j & = n+1, \ldots, n+m
\end{array}
\right\}
\tag{2.13}
$$

Herein $\tau(\rho) \neq \rho$ denotes the unique number, such that $f(\rho) = f(\tau(\rho))$ for all $\rho \neq \sigma$, cf. figure 2.1.

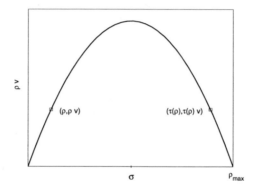

Figure 2.1: A sample fundamental diagram for the LWR–model. $0 \leq \sigma \leq \rho_{max}$ is the point at which the fundamental diagram attains its maximum.

A matrix $\mathcal{A} = (\alpha_{kj}) \in \mathbb{R}^{m \times n}$ is introduced in [19]. This matrix depends on the junction. $\mathcal{A}$ predicts the wishes of the drivers: α_{kj} describes the percentages of drives who want to drive from road j to road k. We assume that $\mathcal{A}$ satisfies ($j', j = 1, \ldots, n$)

$$
\left.
\begin{array}{llll}
\alpha_{kj} \neq \alpha_{kj'} & \forall j \neq j', & k = n+1, \ldots, n+m \\
0 < \alpha_{kn} < 1 & & k = n+1, \ldots, n+m \\
\displaystyle\sum_{k=n+1}^{n+m} \alpha_{kj} = 1 & \forall j
\end{array}
\right\}
\tag{2.14}
$$

The coupling condition (2.7) then reads

$$
f(\bar{\rho}_k) = \sum_{j=1}^{n} \alpha_{kj} f(\bar{\rho}_j) \qquad k = n+1, \ldots, n+m
\tag{2.15}
$$

16

Furthermore, Coclite/Piccoli introduce a function E, which measures the flux on the ingoing roads on the junction

$$E(\bar{\rho}_1, \ldots, \bar{\rho}_n) = \sum_{j=1}^{n} f(\bar{\rho}_j)$$

Theorem 3.1 in [19] states that the problem

$$\max_{\bar{\rho}_j, j=1,\ldots,n} E \qquad \text{subject to } (2.13), (2.14) \text{ and } (2.15) \qquad (2.16)$$

has a unique weak solution in the sense of definition 2.1.2 if we have at least as many outgoing roads as ingoing roads at a junction, i.e., $|\delta_v^-| \leq |\delta_v^+|$ for $v \in V$.

We outline the solution process at the junction in more detail. Consider a junction $v \in V$ with $|\delta_v^-| \leq |\delta_v^+|$. Assume we have Riemann data $\tilde{\rho}_j$ for $j = 1, \ldots, n + m$. Using (2.13) we can determine the feasible set for $\bar{\rho}_j$ for each ingoing road $j = 1, \ldots, n$ and outgoing road $j = n + 1, \ldots, n + m$. Additionally, we assume that the matrix $\mathcal{A}$ is given and satisfies the conditions (2.14). Then we can solve (2.16) and –as shown in [19]– obtain *unique* values $\bar{\rho}_j, j = 1, \ldots, n + m$.

The solution to the Riemann problem at a junction is then constructed as follows. We consider the Riemann problem (2.11) on *ingoing roads*. For the given Riemann data $\tilde{\rho}_j$ and the computed value $\bar{\rho}_j$ the entropic solution to (2.11) is either constant w.r.t. time or has a negative speed of propagation. I.e., the solution travels away from the node $v \in V$. For an *outgoing road* $j = n + 1, \ldots, n + m$ we consider the Riemann problem (2.12). The entropic solution to (2.12) for the given Riemann data $\tilde{\rho}_j$ and the computed value $\bar{\rho}_j$ is either constant w.r.t. time or has a positive speed of propagation. I.e., the solution travels away from the node $v \in V$.

The feasible regions for $\bar{\rho}_j$ (given by (2.13)) depend on the given initial data $\tilde{\rho}_j$. Therefore, the solution to (2.16) depends on the given initial data. Conditions (2.13) ensure that the entropic solution to Riemann problems travels in the correct direction depending on the type of road under consideration.

Remark 2.1.4. *If we have more ingoing roads than outgoing ones, suitable additional conditions have to be added to the optimization formulation (2.16) to ensure uniqueness of the optimal solution $\bar{\rho}$. So–called "equal priority rules" [48, 54] are a possible choice for these conditions, cf. (2.104). The solution for the Riemann problem at a junction is then constructed analogously to the outlined procedure.*

With these preliminaries we can define a weak solution on the network and prove its existence and uniqueness, cf. definition 2.1, definition 2.2, as well as the theorems in sections 4–8 in [19].

Definition 2.1.5. Weak solution of the network problem *Assume that f is smooth and concave with $f(0) = f(\rho_{max}) = 0$. Given an arbitrary network with roads $j \in A$ and junctions $v \in V$ and initial data $\rho_{j,0} \in L^\infty([a_j, b_j]) \, \forall j \in A$ with bounded variation, we define a solution $\rho_j(x,t) \, \forall j \in A$, $x \in [a_j, b_j]$, $t \in \mathbb{R}^+$ to the problem (2.4) and (2.91) as follows:*

17

Let ϕ_j, $j \in A$ be a family of smooth test functions with compact support in $[a_j, b_j]$ which are smooth accross the junctions in the sense of (2.9). A family of functions $\rho_j \in C(L^1_{loc}([a_j, b_j]), \mathbb{R}^+)$, $j \in A$ is called weak solution of the network problem, iff it satisfies (2.10). Furthermore, if ρ_j has bounded variation

$$f(\rho_j(a_j^+, \cdot)) = \sum_{l=1}^{n} \alpha_{jl} f(\rho_l(b_l^-, \cdot)) \quad \forall j = n+1, \ldots, n+m \tag{2.17}$$

$$\sum_{j=1}^{n} f(\rho_j(b_j^-, \cdot)) = \sum_{j=n+1}^{n+m} f(\rho_j(a_j^+, \cdot)) \tag{2.18}$$

is maximum subject to (2.17)

holds for each junction $v \in V$, where α_{jl} are the entries of a matrix $\mathcal{A}^k$ satisfying assertion (2.14) with the previous notation of in– and outgoing roads.

Theorem 2.1.6. (Existence and Uniqueness) *Let all assumptions of definition 2.1.5 hold except those on the flux functions. We assume for the flux functions that $f : [0,1] \to \mathbb{R}$ is continuous, stritly concave, $f(0) = f(1) = 0$ and there exists a $\sigma \in (0,1)$ such that f is smooth on $[0, \sigma)$ and on $(\sigma, 1]$ and*

$$0 < c \le f'(x) \le C < \infty \quad \forall x \in [0, \sigma) \cup (\sigma, 1]$$

Consider a road network in which each junction has at most two ingoing and two outgoing roads. Fix $T > 0$. Then there exists a unique solution in the sense of definition 2.1.5 on $[0, T]$.

This weak solution of the network problem is obtained by the Front-Tracking-Method [62, 23].

Unless stated otherwise, we restrict our attention to graphs with junctions of degree three in the sequel. There are two types of junctions for a node $v \in V$ in these graphs: the first junction type has one ingoing street $i = 1$ with end b_1 at the junction and two outgoing streets labeled $i = 2$, $i = 3$ with ends a_2, a_3 at the junction. At such a junction, a time-dependent control $t \to \alpha_v(t)$ is applied. This control distributes the flux among the outgoing roads and can be seen as the percentage of drivers that have to go to road 2 whereas $1 - \alpha_v(t)$ percent of the drivers have to choose road 3. More precisely, the density flux is split up according to $\alpha_v(t)$ and $1 - \alpha_v(t)$, cf. (2.19). At the second type of junction the roads $i = 1$ and $i = 2$ with ends b_1 and b_2 at the junction merge to road 3 with end a_3 at the junction and the traffic flow is not controlled, since it is determined uniquely by the conservation of cars, see Figure 2.2.

Under the above assumptions the complete coupling conditions [19] read

A1 Boundary values for a junction $v \in V$ of the First Type

$$f_2(\bar{\rho}_2(a_2, t)) = \alpha_v(t) f_1(\bar{\rho}_1(b_1, t)), \tag{2.19a}$$
$$f_3(\bar{\rho}_s(a_3, t)) = (1 - \alpha_v(t)) f_1(\bar{\rho}_1(b_1, t)) \tag{2.19b}$$

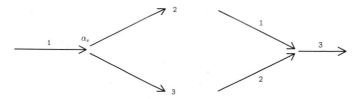

Figure 2.2: Junction at a node $v \in V$ of the first type (left) with $\delta_v^- = \{1\}, \delta_v^+ = \{2, 3\}$ and control $\alpha_v(t)$ and second type (right) with $\delta_v^- = \{1, 2\}$ and $\delta_v^+ = \{3\}$, respectively.

A2 Boundary values for a junction of the Second Type

$$f_3(\bar{\rho}_3(a_3, t)) = f_1(\bar{\rho}_1(b_1, t)) + f_2(\bar{\rho}_2(b_2, t)) \tag{2.20}$$

A3 Boundary values for a road $j \in A$ incoming to the network:

$$\rho_j(a_j, t) = u_j(t) \tag{2.21}$$

Summarizing, the reference model on a network $G = (V, A)$ reads

$$\partial_t \rho_j + \partial_x f_j(\rho_j) = 0 \qquad \forall (x, t) \in [a_j, b_j] \times [0, T] \tag{2.22a}$$
$$\rho_j(x, 0) = \rho_{j,0}(x) \qquad \forall x \in [a_j, b_j] \tag{2.22b}$$

with boundary values given by $(2.19), (2.20)$ and (2.21), respectively $\tag{2.22c}$

2.1.2 Extension to multi–commodity road networks

In the sequel we will discuss various extensions to the reference network model to incorporate driver's source–destination preferences which are not accounted for in the model (2.22).

In [34] a first approach into this direction was presented, but it deviates slightly from the one presented here. Furthermore, the material below differs from [34] in the following way: their work can be seen as an excellent theoretical study of the introduced model, whereas we emphasize simulation and optimization aspects, see below.

We note that the extension of single–class models to multi–class models is an active area of research. A corresponding extension for the Aw–Rascle model [7] is discussed in [8]. Other articles concerning multi–class extensions of the LWR–model are [105, 107]. Multi–population models are addressed in [10, 108].

Network formulation of a multicommodity approach

We consider a road network. Recall that in this text we model the former as a finite, directed graph $G = (V, A)$, where each arc $j \in A$ models a uni-directional road and each vertex $v \in V$ a junction. Furthermore, each road j is identified by an interval $[a_j, b_j]$ where we allow a_j or b_j to be infinity.

We assume that we are given a set $\mathcal{S} \subset V$ of *source* or starting nodes and a set $\mathcal{D}$ of *destination* or terminal nodes. We assume w.l.o.g. $\mathcal{S} \cap \mathcal{D} = \emptyset$ (cf. remark 2.1.7 below) and introduce the set of all source–destination pairings $\mathcal{M}$

$$V \times V \supset \mathcal{M} := \mathcal{S} \times \mathcal{D} \qquad I := |\mathcal{M}| \tag{2.23}$$

In the sequel we will define I commodities. For every $(s_i, d_i) \in \mathcal{M}$ we set

$$\vec{\gamma}^{\,i}(t, x) \;:=\; (\gamma_1^i(t, x), \dots, \gamma_{|E|}^i(t, x)) \qquad i = 1, \dots, I \tag{2.24}$$

$\gamma_j^i(t, x)$ denotes the *percentage* of drivers of commodity i on road j at point x at time t. Thus, in our notation the index i codes the source-destination relationship and the index j holds the information concerning the road.

Remark 2.1.7. *If $\mathcal{S} \cap \mathcal{D} = \mathcal{C} \neq \emptyset$, we proceed as follows. For $c \in \mathcal{C}$ we take $(s_i, c), (c, d_j) \in \mathcal{M}$. We add a node v to the set of vertices V and add the edge (c, v) to the edge set E; this new road is assigned the length $\varepsilon > 0$ and a capacity that is large enough to prevent traffic jams. Then we set $\mathcal{D} = (\mathcal{D} \cup v) \setminus c$. Now we have $\mathcal{S} \cap \mathcal{D} = \mathcal{C} \setminus \{c\}$. We can apply this procedure recursively and finally achieve $\mathcal{S} \cap \mathcal{D} = \emptyset$.*

Of course, we have to describe how the commodities $\vec{\gamma}^{\,i}$ are related to each other and how they evolve over time. We assume we are given a velocity distribution $v_j \geq 0, j = 1, \dots, |E|$, i.e., we know the average velocity on every road in the network. Then the property γ_j^i should be advected with velocity v_j for every $i = 1, \dots, I$:

$$\partial_t \gamma_j^i(x, t) + v_j(x, t) \cdot \partial_x \gamma_j^i(x, t) \;=\; 0, \qquad i = 1, \dots, I, \; j = 1, \dots, |A| \tag{2.25}$$

For every i, j (2.25) is a hyperbolic equation in *nonconservative* form. Since γ_j^i models the percentage of drivers, we need to impose the following restrictions

$$0 \leq \gamma_j^i(x, t) \leq 1 \qquad \forall i, j, (x, t) \in [a_j, b_j] \times [0, T] \tag{2.26}$$

$$\sum_{i \in I} \gamma_j^i(x, t) \;=\; 1 \qquad \forall j \in \{1, \dots, |A|\} \tag{2.27}$$

In order to obtain a well-posed problem, we need to provide an initial profile at $t = 0$ and for roads ingoing to the network we need an inflow profile over time:

$$\gamma_j^i(x, 0) \;=\; g_{ij}(x) \qquad i = 1, \dots, I, \quad j = 1, \dots, |A| \tag{2.28}$$

$$\gamma_j^i(0, t) \;=\; h_{ij}(t) \qquad i = 1, \dots, I, \quad j \text{ is an ingoing road} \tag{2.29}$$

Here, $g_{ij}(x)$ defines the initial profile on the network and $h_{ij}(t)$ denotes the inflow profile over time for an ingoing road j. Note, that by virtue of (2.26) and (2.27) we require

$$0 \leq g_{ij}(x) \leq 1$$
$$0 \leq h_{ij}(t) \leq 1$$
$$\sum_{i \in I} g_{ij}(x) \;=\; 1$$
$$\sum_{i \in I} h_{ij}(t) \;=\; 1$$

Remark 2.1.8. *The velocities v_j are assumed to be given. Of course, they can be obtained as solution of the LWR- or the AR-equations ([85] or [7], respectively) or any other traffic flow model.*

In order to obtain a network model, we need to take care of the coupling at junctions. Our approach is based on an idea presented in [34]. At a junction $v \in V$ with $n > 1$ outgoing roads, they introduce the control variables $\alpha_{hl}(t)$. α_{hl} denotes the amount of drivers that are directed from ingoing road l to outgoing road h at time t at a junction v. In order to ensure flow-conservation, one requires the controls to satisfy

$$\sum_{\{h | h \in \delta_v^+\}} \alpha_{hl}(t) = 1 \qquad l \in \delta_v^- \tag{2.30}$$

Since we have different commodities, we expect to have controls corresponding to these commodities at a junction $v \in V$. We investigate the general case of m ingoing and n outgoing roads first and then restrict ourselves to a more detailed discussion of the two cases $m = 1, n = 2$ (dispersing junction) and $m = 2, n = 1$ (merging junction).

Let δ_v^- denote the set of ingoing roads at junction v and δ_v^+ the set of outgoing roads at junction v. According to our assumption, we have

$$\delta_v^- = \{1, \dots, \varepsilon, \dots, m\} \qquad \delta_v^+ = \{1, \dots, \nu, \dots, n\} \tag{2.31}$$

Remark 2.1.9. *As indicated in figure 2.3 the numbering in (2.31) is local only. For a given network topology we have a bijective mapping from $\{1, \dots, |A|\}$ to A. Additionally, at every junction $v \in V$ we have an injective map $\{1, \dots, m, m+1, \dots, m+n\} \to \{1, \dots, |A|\}$ where $m = |\delta_v^-|$ and $n = |\delta_v^-|$.*

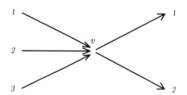

Figure 2.3: Situation at a junction $v \in V$ with $m = 3$ ingoing and $n = 2$ outgoing edges (roads)

We introduce for every commodity $i = 1, \dots, I$ at the junction v $m \cdot n$ control variables:

$$\alpha_{\varepsilon\nu}^i(v; t) \qquad \varepsilon \in \delta_v^-, \nu \in \delta_v^+ \tag{2.32}$$

The interpretation is straightforward: $\alpha_{\varepsilon\nu}^i(v; t)$ describes for commodity i at time t the amount of ingoing flow on road ε that is directed to outgoing road ν at junction $v \in V$.

Of course, we again have some constraints on the choice of the controls. We require

$$0 \le \alpha_{\varepsilon\nu}^i \le 1 \qquad \forall i, \varepsilon, \nu \tag{2.33}$$

$$\sum_{\nu=1}^n \alpha_{\varepsilon\nu}^i = 1 \qquad \forall \varepsilon = 1, \dots, m, \ i = 1, \dots, I \tag{2.34}$$

21

(2.34) resembles the condition (2.30); we need it for every commodity $i \in I$ to ensure flow conservation at a junction for every commodity. With the aid of the control variables (2.32), we define a control $\alpha_{\varepsilon\nu}(v;t)$ that is independent of the commodity. Additionally, we propose coupling conditions for the commodities $\vec{\gamma}^i$:

$$\alpha_{\varepsilon\nu}(v;t) \quad := \quad \frac{\sum\limits_{i=1}^{I} \alpha_{\varepsilon\nu}^i(v;t)\gamma_\varepsilon^i(b_\varepsilon)}{\sum\limits_{i=1}^{I} \gamma_\varepsilon^i(b_\varepsilon)} \quad \overset{(2.27)}{=} \quad \sum_{i=1}^{I} \alpha_{\varepsilon\nu}^i(v;t)\gamma_\varepsilon^i(b_\varepsilon) \tag{2.35}$$

$$\gamma_\nu^i(a_\nu) \quad := \quad \frac{\sum\limits_{\varepsilon=1}^{m} \alpha_{\varepsilon\nu}^i \gamma_\varepsilon^i(b_\varepsilon) f_\varepsilon(\rho_\varepsilon(b_\varepsilon,t))}{\sum\limits_{i=1}^{I}\sum\limits_{\varepsilon=1}^{m} \alpha_{\varepsilon\nu}^i \gamma_\varepsilon^i(b_\varepsilon) f_\varepsilon(\rho_\varepsilon(b_\varepsilon,t))} \quad \nu = 1,\ldots,n \tag{2.36}$$

The control $\alpha_{\varepsilon\nu}(v;t)$ corresponds to the control $\alpha_{hl}(v;t)$ in [19]. We have $0 \le \alpha_{\varepsilon\nu}(v;t) \le 1$ and a simple computation yields

$$\sum_{\nu=1}^{n} \alpha_{\varepsilon\nu}(v;t) \quad = \quad 1$$

For a given traffic flow model, we have a coupling at a junction v in terms of the fluxes and the control variables $\alpha_{\varepsilon\nu}(v;t)$; the controls $\alpha_{\varepsilon\nu}^i(v;t)$ enter in the model by formula (2.35) as well as the terminal values of the commodity equation.

Furthermore, we provide a formula for the computation of an initial value for the commodity equation via (2.36). This definition is in accordance with condition (2.27), as we obtain

$$\sum_{i=1}^{I} \gamma_\nu^i(a_\nu) \quad \overset{(2.36)}{=} \quad \sum_{i=1}^{I} \frac{\sum\limits_{\varepsilon=1}^{m} \alpha_{\varepsilon\nu}^i \gamma_\varepsilon^i(b_\varepsilon) f_\varepsilon(\rho_\varepsilon(b_\varepsilon,t))}{\sum\limits_{i=1}^{I}\sum\limits_{\varepsilon=1}^{m} \alpha_{\varepsilon\nu}^i \gamma_\varepsilon^i(b_\varepsilon) f_\varepsilon(\rho_\varepsilon(b_\varepsilon,t))}$$

$$= \quad 1$$

Example 2.1.10. *For the case of a merging junction ($m = 2, n = 1$) and 2 commodities on the ingoing roads, we obtain*

$$\alpha_{\varepsilon 1}^i \quad = \quad 1 \quad \varepsilon = 1,2, \; i = 1,2 \quad \text{by (2.34)}$$

$$\gamma_3^i(a_3) \quad = \quad \frac{\gamma_1^i(b_1)f_1(\rho_1(b_1,t)) + \gamma_2^i(b_2)f_2(\rho_2(b_2,t))}{f_1(\rho_1(b_1,t)) + f_2(\rho_2(b_2,t))} \quad \text{by (2.36)}$$

$$= \quad \mu_1\,\gamma_1^i(b_1) + \mu_2\,\gamma_2^i(b_2) \tag{2.37}$$

$$\mu_1 \quad = \quad \frac{f_1(\rho_1(b_1,t))}{f_1(\rho_1(b_1,t)) + f_2(\rho_2(b_2,t))}$$

$$\mu_2 \quad = \quad \frac{f_2(\rho_2(b_2,t))}{f_1(\rho_1(b_1,t)) + f_2(\rho_2(b_2,t))}$$

$$\alpha_{11} \quad = \quad 1$$

$$\alpha_{21} \quad = \quad 1$$

where we have used (2.27). For a flux function f the condition of flow conservation at a merging junction reads in the approach without any commodities

$$f_3(\rho_3(a_3, t)) = f_1(\rho_1(b_1, t)) + f_2(\rho_2(b_2, t))$$

If we now set

$$f_\nu(\rho_\nu(a_\nu, t)) = \sum_{\varepsilon=1}^{m} \alpha_{\varepsilon\nu}(v; t) f_\varepsilon(\rho_\varepsilon(b_\varepsilon, t))$$

we exactly recover this formula. In particular, we satisfy the Rankine–Hugoniot condition.

Consider a dispersing junction, i.e., $m = 1$ and $n = 2$. Furthermore, we assume that the outgoing label $\nu = 1$ corresponds to road number 2 and $\nu = 2$ corresponds to road number 3 (in general $\nu = l$ corresponds to road number $m + l$ at a junction $v \in V$ with $|\delta_v^-| = m$ and $|\delta_v^+| = n$). For the case of 2 commodities on the ingoing road, we obtain with our general formulas (note that $f_1(\rho_1(b_1, t))$ cancels in the computation for $\gamma_2^i(a_2)$ and $\gamma_3^i(a_3)$)

$$\alpha_{12}^i + \alpha_{13}^i = 1 \quad i = 1, 2 \quad \text{by (2.34)}$$

$$\gamma_2^i(a_2) = \frac{\alpha_{12}^i \gamma_1^i(b_1)}{\sum\limits_{i=1}^{2} \alpha_{12}^i \gamma_1^i(b_1)} \quad i = 1, 2 \quad \text{by (2.36)}$$

$$\gamma_3^i(a_3) = \frac{\alpha_{13}^i \gamma_1^i(b_1)}{\sum\limits_{i=1}^{2} \alpha_{13}^i \gamma_1^i(b_1)} \quad i = 1, 2 \quad \text{by (2.36)}$$

$$\alpha_{12} = \alpha_{12}^1 \gamma_1^1(b_1) + \alpha_{12}^2 \gamma_1^2(b_1) \tag{2.38}$$

$$\alpha_{13} = \alpha_{13}^1 \gamma_1^1(b_1) + \alpha_{13}^2 \gamma_1^2(b_1) \tag{2.39}$$

Note that the definition of the coupling condition

$$f_\nu(\rho_\nu(a_\nu, t)) := \sum_{\varepsilon=1}^{m} \alpha_{\varepsilon\nu}(v; t) f_\varepsilon(\rho_\varepsilon(b_\varepsilon, t)) \quad \nu = 1, \ldots, m$$

is consistent with the approach taken in [19]. In particular, the results concerning uniqueness of solutions of the maximization problems at junctions (cf. (2.16) and remark 2.1.4) is ensured by the modified coupling conditions which now include commodities. Furthermore, we recover the reference model formulas for the coupling conditions if we have only *one* commodity.

On the determination of $\alpha_{\varepsilon\nu}^i(v; t)$

In the following we will supress the time–dependence of the functions $\alpha_{\varepsilon\nu}^i(v; t)$ and refer to the *variable* $\alpha_{\varepsilon\nu}^i(v)$. In fact, in numerical simulations we will have a time–discretization and we will assume piecewise constant controls. Therefore, we really have variables $\alpha_{\varepsilon\nu}^{i,j}(v) = \alpha_{\varepsilon\nu}^i(v; t_j)$.

As before, at a given junction $v \in V$ the variable $\alpha^i_{\varepsilon\nu}(v), \varepsilon \in \delta^-_v, \nu \in \delta^+_v, i = 1, \dots, I$ denotes the amount of drivers of commodity i that are distributed from ingoing road ε to outgoing road ν. However, the introduction of the variables $\alpha^i_{\varepsilon\nu}(v)$ leads to some complications. One of the issues to be treated concerns the sheer amount of control variables introduced in this formulation. As we will see below, some of the variables need to be fixed in order to obtain a well–defined model and allow for a unique solution to an associated optimization problem. In appendix B we describe an algorithm which –prior to a simulation or optimization procedure– automatically determines controls which need to be set and initializes them appropriately.

For a given network, the value of some of the variables $\alpha^i_{\varepsilon\nu}(v)$ is determined by the network topology. We want drivers to reach certain destinations and this naturally translates into restrictions on the controls; these restrictions will be made precise below. Without them the model will not work correctly, i.e., we can not ensure that the drivers' preferences are respected in our model. In particular, for a commodity $i \in I$, the paths leading from the source s_i to destination d_i will play a vital role in the process of determining the value of a variable $\alpha^i_{\varepsilon\nu}(v; t)$. Therefore, we define the set $\mathcal{P}_i$ as the set of *all* paths leading from s_i to d_i:

$$
\begin{aligned}
\mathcal{P}_i \;\; := \;\; \{ P_z \mid & P_z = (s_i, v_1, \dots, v_z, d_i), v_j \in V, \\
& (s_i, v_1), (v_z, d_i), (v_p, v_{p+1}) \in A, p = 1, \dots, z - 1 \}
\end{aligned} \tag{2.40}
$$

Now we can precisely state when a road $j \in A$ belongs to a path P_z, i.e., $j \in P_z$. This is not immediate since $P_z \subset V^{z+2}$ and $\varepsilon \in A \subset V \times V$.

Definition 2.1.11. *Consider some $j \in A$. We say that road $j = (v, w) \subset V \times V$ belongs to a path P_z iff there exists an index $p \in \{1, \dots, z-1\}$ such that*

$$
(v, w) = (v_p, v_{p+1}) \quad \text{or} \quad (v, w) = (s_i, v_1) \quad \text{or} \quad (v, w) = (v_z, d_i)
$$

In order to fix some of the controls $\alpha^i_{\varepsilon\nu}(v)$, $v \in V, \varepsilon \in \delta^-_v, \nu \in \delta^+_v$, we distinguish the following four cases at a junction $v \in V$ for every commodity $i = 1, \dots, I$, ignoring the constraint (2.34) for now:

1. $\varepsilon \in P_z$ for at least one $P_z \in \mathcal{P}_i$, $\nu \notin P_z$ for all $P_z \in \mathcal{P}_i$
   In this case, the ingoing road can have a nonzero commodity value (i.e. $\gamma^i_\varepsilon(b_\varepsilon, t) > 0$), but we can not reach destination d_i via road ν. Consequently, we need

   $$
   \alpha^i_{\varepsilon\nu}(v; t) \;\; := \;\; 0
   $$

   Of course, if $\gamma^i_\varepsilon(b_\varepsilon, t) > 0$ for some time t, we run into trouble, since the destination d_i can not be reached. Recall, that we just want to fix or eliminate some variables at this stage and are not concerned with solving equations (2.25) correctly.

2. $\varepsilon \in P_z$ for at least one $P_z \in \mathcal{P}_i$, $\nu \in P_z$ for at least one $P_z \in \mathcal{P}_i$
   Here we can have a nonzero commodity value (i.e. $\gamma^i_\varepsilon(b_\varepsilon) > 0$) and we can reach destination d_i via road ν. Hence we can not fix $\alpha^i_{\varepsilon\nu}$ here and we need to look at (2.34), which will be done below. In particular we note, that the choice of $\alpha^i_{\varepsilon\nu}$ is not arbitrary.

24

3. $\varepsilon \notin P_z$ for all $P_z \in \mathcal{P}_i$, $\nu \in P_z$ for at least one $P_z \in \mathcal{P}_i$

   In this case, a control variable $\alpha^i_{\varepsilon\nu}(v;t)$ does not exist, since there is no chance that we will direct a portion of some γ^i_j on road ε. We will only distribute at a junction v on those roads, that are member of a path from s_i to d_i. In particular, we do not have to satisfy (2.34) in this case. However, for notational and computational convenience, we set $\gamma^i_\varepsilon(b_\varepsilon) = 0$ and $\alpha^i_{\varepsilon\nu}(v) = 0$.

4. $\varepsilon \notin P_z$ for all $P_z \in \mathcal{P}_i$, $\nu \notin P_z$ for all $P_z \in \mathcal{P}_i$

   This case is similar to the previous one: we set $\gamma^i_\varepsilon(b_\varepsilon) = 0$ and $\alpha^i_{\varepsilon\nu}(v) = 0$ for computational convenience, although these variables are nonexistent in real life (there is nothing to distribute). Again, we do not have to take into account (2.34).

We have to incorporate (2.34) into the considerations. As we have observed, the only interesting case is $\varepsilon \in P_z$ for at least one $P_z \in \mathcal{P}_i$. Then (2.34) reads, using cases 1 and 2

$$
\begin{aligned}
1 &= \sum_{\nu=1}^n \alpha^i_{\varepsilon\nu}(v;t) \\
&= \sum_{\substack{\nu \in P_z \text{ for some } P_z \in \mathcal{P}_i}} \alpha^i_{\varepsilon\nu}(v;t) + \sum_{\substack{\nu \notin P_z \text{ for all } P_z \in \mathcal{P}_i}} \alpha^i_{\varepsilon\nu}(v;t) \\
&= \sum_{\substack{\nu \in P_z \text{ for some } P_z \in \mathcal{P}_i}} \alpha^i_{\varepsilon\nu}(v;t)
\end{aligned}
$$

What is written here is vital for an optimization routine to run correctly; it defines the degree of freedom at a node $v \in V$. We can choose the values $\alpha^i_{\varepsilon\nu}(v;t), \varepsilon \in \delta^-_v, \nu \in \delta^+_v$, $\varepsilon, \nu \in P_z$ for some $P_z \in \mathcal{P}_i$ such that

$$
\sum_{\substack{\nu \in P_z \text{ for some } P_z \in \mathcal{P}_i}} \alpha^i_{\varepsilon\nu}(v;t) = 1 \tag{2.41}
$$

Summarizing, we have isolated the interesting variables:

$$
\alpha^i_{\varepsilon\nu} \quad \varepsilon \in \delta^-_v, \nu \in \delta^+_v, \varepsilon, \nu \in P_z \text{ for some } P_z \in \mathcal{P}_i \tag{2.42}
$$

If $\varepsilon \in P_z$ for some $P_z \in \mathcal{P}_i$ the case $\{\nu \in P_z \text{ for some } P_z \in \mathcal{P}_i\} = \emptyset$ must not occur, since that would indicate that we have some inflow on road ε for commodity i, but there is no way to get to the destination d_i from junction $v \in V$. For $\varepsilon \notin P_z$ for all $P_z \in \mathcal{P}_i$, (2.34) is irrelevant.

Another crucial aspect is that we can determine the variables (2.42) in a preprocessing step that has to be done only once for a given network topology, cf. appendix B. This makes the whole model more practical.

Example 2.1.12. *In order to illuminate the presented ideas we consider the simplest example for an intersection inside a network. It consists of one incoming and two outgoing roads, labeled 1 and 2, 3, respectively (cf. example 2.1.10).*

We assume that we have $I = 2$ commodities. The first commodity can reach its destination d_1 via both outgoing roads 2 and 3, whereas we assume that the second *commodity can only take road 3 to get to d_2. In the multi–commodity model we have four variable controls, cf. (2.38) and (2.39):*

$$\vec{\alpha} \quad := \quad (\alpha_{12}^1, \alpha_{13}^1, \alpha_{12}^2, \alpha_{13}^2)$$

By our assumption in this example and case 1 discussed above we have $\alpha_{12}^2 = 0$ and then we obtain $\alpha_{13}^2 = 1$ since we require $\alpha_{12}^2 + \alpha_{13}^2 = 1$ as in example 2.1.10. For the first commodity we need to apply case 2 and hence we are led to

$$\alpha_{12}^1 + \alpha_{13}^1 \quad = \quad 1$$

In particular, we have reduced the number of free controls at this node from four to one and can only choose one control freely at this node. For the model to work correctly, the applicable control vector reads

$$\vec{\alpha} = (\alpha_{12}^1, 1 - \alpha_{12}^1, 0, 1), \qquad 0 \le \alpha_{12}^1 \le 1$$

For a simulation corresponding to a particular control we choose values for the variable coefficients $\alpha_{\varepsilon\nu}^i(v)$ given by (2.41). Then the controls $\alpha_{\varepsilon\nu}$ are determined (cf. (2.35)) and we can use these controls to solve the optimization problems at junctions (cf. (2.16) and remark 2.1.4) to solve certain (half)–Riemann problems as described for the reference network [19] in section 2.1.1. This guarantees that we compute the correct weak network solution, i.e., the solution satisfying definition 2.1.5.

Summarizing, the multicommodity model with $I = |\mathcal{M}|$ commodities on a road network $G = (V, A)$ can be stated as follows for all $v \in V$ and $i = 1, \ldots, I$

$$\partial_t \rho_j + \partial_x f(\rho_j) = 0, \qquad j = 1, \ldots, |A| \tag{2.43a}$$

$$\partial_t \gamma_j^i(x, t) + v_j(x, t) \cdot \partial_x \gamma_j^i(x, t) = 0, \qquad j = 1, \ldots, |A| \tag{2.43b}$$

$$\rho_j(x, 0) = \tilde{g}_j(x) \qquad j = 1, \ldots, |A| \tag{2.43c}$$

$$\gamma_j^i(x, 0) = g_{ij}(x) \qquad j = 1, \ldots, |A| \tag{2.43d}$$

$$\rho_j(0, t) = \tilde{h}_j(t) \qquad j \text{ is an ingoing road} \tag{2.43e}$$

$$\gamma_j^i(0, t) = h_{ij}(t) \qquad j \text{ is an ingoing road} \tag{2.43f}$$

$$f_\nu(a_\nu, t) = \sum_{\varepsilon \in \delta_v^-} \alpha_{\varepsilon\nu}(v; t) f_\varepsilon(b_\varepsilon, t), \qquad \nu \in \delta_v^+ \tag{2.43g}$$

$$\alpha_{\varepsilon\nu}(v; t) = \sum_{i=1}^I \alpha_{\varepsilon\nu}^i(v; t) \gamma_\varepsilon^i(b_\varepsilon), \qquad \varepsilon \in \delta_v^-, \nu \in \delta_v^+ \tag{2.43h}$$

$$\gamma_\nu^i(a_\nu) = \frac{\sum\limits_{\varepsilon \in \delta_v^-} \alpha_{\varepsilon\nu}^i \gamma_\varepsilon^i(b_\varepsilon) f_\varepsilon(\rho_\varepsilon(b_\varepsilon, t))}{\sum\limits_{i=1}^I \sum\limits_{\varepsilon \in \delta_v^-} \alpha_{\varepsilon\nu}^i \gamma_\varepsilon^i(b_\varepsilon) f_\varepsilon(\rho_\varepsilon(b_\varepsilon, t))} \qquad \nu \in \delta_v^+, \tag{2.43i}$$

$$0 \le \alpha_{\varepsilon,\nu}^i \le 1 \qquad \varepsilon \in \delta_v^-, \nu \in \delta_v^+ \tag{2.43j}$$

$$\sum_{\nu \in \delta_v^+} \alpha_{\varepsilon,\nu}^i = 1 \qquad \forall \varepsilon \in \delta_v^- \tag{2.43k}$$

2.1.3 Solution to the multi–commodity equation

Some nonconservative hyperbolic equations solely admit measure solutions which can in general *not* be represented by weak solutions [14, 93]. Therefore, we discuss the solution structure of the commodity equation below and include a more thorough discussion in appendix A. Furthermore, we believe the discussion will help to develop a simplified multi–commodity model based on the Front–Tracking method [63, 15].

In [34] one finds the commodity equation (2.25)

$$\partial_t \gamma^i(x,t) + v(x,t) \cdot \partial_x \gamma^i(x,t) = 0 \qquad (2.44)$$

(2.44) constitutes a hyperbolic partial differential equation in nonconservative form. In the sequel we will discuss solutions to this equation. We will drop the index j for the roads in the following as we have done already in (2.44). Additionally we assume that $v(x,t)$ is known and bounded (in our application, we have $v \in L^\infty_{loc}(\mathbb{R} \times [0,T])$).

As we will see below, the coefficient v can be discontinuous in our model and therefore we can not use standard solution techniques. We need to introduce so–called generalized or Filippov characteristics [31] to extend the method of characteristics to these more general equations. In one space–dimension, equation (2.44) can be reformulated in conservative form by differentiation w.r.t. x

$$\partial_t \mu + \partial_x(v\,\mu) = 0 \qquad (2.45)$$

The solution $\mu := \partial_x \gamma$ to (2.45) is in general solely a measure–solution. Additionally, a priori a suitable definition of the product $v\,\mu = v\partial_x\gamma$ is not obvious. If we do not assume an *a priori* relation between v and μ, there are two ways to interpret the product [14]. An approach for the characterization of measure solutions to transport equations can be found in [93]. It coincides with one of the methods presented in [14].

As shown in [93], if the Filippov characteristics are unique one can obtain a unique solution to equation (2.45). For our particular set of equations we can guarantee uniqueness of the Filippov characteristics. Additionally, we can always construct a weak solution in our application. For illustration purposes we describe in appendix A how one can cope with the case in which the Filippov characteristics are not unique. This procedure is suggested in [14] and incorporates the notion of *reversible* and *duality* solution.

In the sequel we outline the main ideas used in the construction of a solution to both (2.44) and (2.45). Frequently we will use figures to illustrate the suggested procedure. For rigorous results concerning existence, uniqueness and stability of solutions to linear transport equations in the presence of a possibly discontinuous coefficent we refer the reader to [14, 93] and the references therein.

Before we review the concept of Filippov characteristics [31] we recall a general solution technique for quasi–linear equations.

Method of characteristics

We briefly review the method of characteristics, a standard technique used to solve quasi–linear equations. We restrict ourselves to the case of two independent variables. As pointed out in [68], we can obtain a solution to equation (2.44) from the solution of the following system of ODEs

$$\frac{dx}{dt} = v(y, x) \tag{2.46a}$$

$$\frac{dy}{dt} = 1 \tag{2.46b}$$

$$\frac{dz}{dt} = 0 \tag{2.46c}$$

To obtain a unique solution to (2.46), we impose the following initial conditions for an arbitrary $\bar{x} \in \mathbb{R}$

$$x(0) = \bar{x}, \quad y(0) = 0, \quad z(x(0), y(0)) = h(x(0)) \tag{2.47}$$

Obviously, in view of (2.47) the solution to (2.46b) and (2.46c) is given by

$$y(t) = t$$
$$z(x, t) = const = h(x(0))$$

If v is continuous and in particular constant $v(y, x) \equiv v_0$, the solution to (2.46a), (2.47) is

$$x(t) = v_0 t + \bar{x}$$

Therefore, the integral surface corresponding to (2.44) with a continuous coefficient v is given by

$$z(x, t) = h(x - v_0 t)$$

This states that we transport the initial value $h(\bar{x})$ along the so–called *characteristic curve* $x(t) = v_0 t + \bar{x}$ that leads through $\bar{x}$, cf. figure 2.4. Additionally, we can immediately verify that the function $\gamma^i(x, t) = z(x, t) = h(x - v_0 t)$ satisfies the partial differential equation (2.44) if h posesses enough regulartiy.

Since v can be discontinuous in our application, i.e.,

$$v(t, x) = \begin{cases} v_l & x - st < x_0 \\ v_r & x - st > x_0 \end{cases} \tag{2.48}$$

for some $s \in \mathbb{R}$ we need a generalized notion of solution of an equation of the form (2.46a). In this context the aforementioned generalized characteristics enter the scene. They facilitate the construction of solutions to problems in which the transport coefficient is discontinuous. A prototype of Filippov characteristics is depicted in figure 2.5.

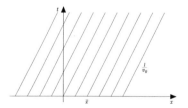
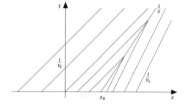

Figure 2.4: Characteristics in $x - t$ plane for constant coefficient $v \equiv v_0$.

Figure 2.5: Generalized characteristics in $x - t$ plane for a discontinuous coefficient.

Filippov characteristics

Filippov characteristics are a key concept to solve hyperbolic equations with discontinuous coefficients. They generalize the well-known method of characteristics (cf. e.g. [68]) for solving a quasi–linear equation. A Filippov characteristic $X(t)$ has to satisfy the differential inclusion

$$\dot{X}(t) \in F(t, x)$$

where $F(t, x)$ has to be suitably defined. There are various ways to define $F(t, x)$, cf. Chapter 2, paragraph 4 in [31]. We will stick to the simplest (and for our purposes sufficient) definition, the convex one.

We consider an equation

$$\dot{x} = f(t, x)$$

with a piecewise continuous function f in a domain G. M is a set (of measure zero) of points of discontinuity of the function f. If at the point (t, x) the function f is continuous, the set $F(t, x)$ consists of one point which coincides with the value of the function f at this point. If the point $(t, x) \in M$ lies on the boundaries of the cross-sections of two domains G_1, G_2 intersected by the plane $t = const$, the set $F(t, x)$ is a segment with edges $f_i(t, x)$, where

$$f_i(t, x) = \lim_{(t,x^*) \in G_i, x^* \to x} f(t, x^*) \tag{2.49}$$

Thus in our situation, $M = \{(t, x_0 + st) \mid t \geq 0\}$, $G_1 = \{(t, x) \mid x < x_0 + st, t > 0\}$ and $G_2 = \{(t, x) \mid x > x_0 + st, t > 0\}$. Hence if $(t, x) \in M$ the set $F(t, x)$ is given by

$$F(t, x) = [v_r, v_l]$$

Of course, this does not help much in computing the characteristics. In particular, it seems we have a lot of freedom in the choice of characteristics. However, a result in [14] states that if $v(t, x)$ satisfies a one–sided Lipschitz condition, we have uniqueness for the Filippov solutions for the forward problem, i.e., the problem

$$\partial_t X(t; 0, \bar{x}) = v(t, X(t; 0, \bar{x}))$$
$$X(0; 0, \bar{x}) = \bar{x}$$

has a unique solution $X(t; 0, \bar{x})$ for $t > 0$.

For $v \in L^\infty(\mathbb{R} \times [0, T])$ one formulation of the *one–sided Lipschitz condition* (OSLC) reads

$$(v(t, x) - v(t, y))(x - y) \leq \alpha(t)(x - y)^2$$

for some $\alpha \in L^1((0, T))$ and almost every $(t, x, y) \in (0, T) \times \mathbb{R} \times \mathbb{R}$. Theorem 2.2 in [93] restates a result from [31]: if the OSLC is satisfied the Filippov characteristics are unique.

In case $v_l > v_r$ the function v defined by (2.48) satisfies the OSLC, e.g., choose $\alpha \equiv 0$. Hence the Filippov characteristics are unique. This suggests, cf. [14], that we can compute these characteristics by an approximation argument.

29

Computation of Filippov Characteristics

We consider the Riemann problem

$$\partial_t \gamma(x,t) + v(x,t) \cdot \partial_x \gamma(x,t) = 0 \tag{2.50a}$$

$$\gamma(0,x) = \begin{cases} \gamma_l & x < 0 \\ \gamma_r & x > 0 \end{cases} \tag{2.50b}$$

with the discontinuous coefficient v from (2.48). Numerically we solve problem (2.50) by an approximation argument. We smoothen v using a mollifier and compute the limit as $\varepsilon \to 0$. For a smooth $v_\varepsilon(t,x)$ we can solve the characteristic equation

$$\partial_t X(t;0,\bar{x}) = v_\varepsilon(t, X(t;0,\bar{x})) \tag{2.51}$$

$$X(0;0,\bar{x}) = \bar{x}$$

for $t \geq 0$ to obtain $X_\varepsilon(t;0,\bar{x})$.

We use a mollifier based on $arctan(x)$, i.e., we choose as an approximation for $\delta(x)$ the function

$$\delta_\varepsilon(x) = \frac{1}{\pi} \frac{\varepsilon}{\varepsilon^2 + x^2}$$

Then one obtains for $s \in \mathbb{R}$

$$v_\varepsilon(t,x) = \frac{1}{2}(v_r + v_l) + \frac{(v_r - v_l)}{\pi} \cdot arctan\left(\frac{x - x_0 - s \cdot t}{\varepsilon}\right) \tag{2.52}$$

The solution of (2.51) can not be computed analytically for $v_\varepsilon(t,x)$ given by (2.52). Therefore, we use a numerical intergration scheme [12] in order to obtain the solution. We present results solely for one particular choice of parameters.

For $v_l > v_r > 0, x_0 = 0$ the numerically computed, approximate characteristics are shown in figures 2.6 and 2.7 for different values of ϵ. The dash–dotted line corresponds to the shock at which the discontinuity in the coefficient $v(t,x)$ travels. The result indicates that the characteristics converge. Of course we have tested this for different parameter settings.

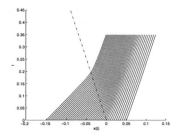

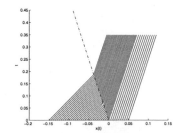

Figure 2.6: Sample characteristics for $\epsilon = 1e^{-2}$ and $s < 0$

Figure 2.7: Sample characteristics for $\epsilon = 1e^{-4}$ and $s < 0$

Remark 2.1.13. *In order to see the 'kinks' we have to resolve ε appropriately, i.e., we need to discretize w.r.t. x such that we have some initial values $\bar{x} \in (x_0 - \varepsilon, x_0 + \varepsilon)$, since outside this interval the characteristics are straight lines with slopes $\frac{1}{v_l}$ and $\frac{1}{v_r}$, respectively.*

Remark 2.1.14. *As a result of a case study (see appendix A) we obtain certain restrictions on admissible relations between s, v_l and v_r for our particular application. These results are vital to apply Front Tracking as solution and simplification technique.*

Construction of weak solutions

We denote the initial datum for the Filippov characterstic by $x(0) = \bar{x}$ and set $s = 0$ in (2.48) for simplicity; the case $s \neq 0$ is treated in appendix A. The point of discontinuity in v is denoted by x_0, cf. (2.48). Then we obtain the following Filippov characteristics:

1. $\bar{x} < x_0$

$$X(t; 0, \bar{x}) = \begin{cases} v_l \cdot t + \bar{x} & t < \frac{x_0 - \bar{x}}{v_l} \\ v_r \cdot \left(t - \frac{x_0 - \bar{x}}{v_l}\right) + x_0 & t \geq \frac{x_0 - \bar{x}}{v_l} \end{cases} \tag{2.53}$$

2. $\bar{x} \geq x_0$

$$X(t; 0, \bar{x}) = v_r \cdot t + \bar{x} \tag{2.54}$$

We illustrate the characteristics in figure 2.8.

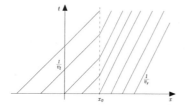
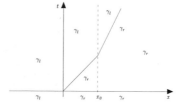

Figure 2.8: Characteristics in $x - t$ plane

Figure 2.9: Distribution of initial values in $x - t$ plane

We found a unique characteristic for a starting point $(t = 0, x = \bar{x})$; classically along characteristics the initial value is just transported. This principle carries over to generalized characteristics (Theorem 3.4 and 3.5 in [14]). Therefore, the solution structure of the Riemann problem (2.50) is the one depicted in figure 2.9. Additionally, the solution of this form is *unique*.

31

Remark 2.1.15. *Note that a weak solution needs to satisfy the following weak formulation of (2.44)*

$$\int\limits_{-\infty}^{\infty} \int\limits_{0}^{\infty} a^i(\phi_t + v\phi_x)\, dt\, dx \;=\; -\int\limits_{-\infty}^{\infty} a^i(0,x)\phi(0,x)\, dx$$

$$+(v_l - v_r)\int\limits_{0}^{\infty} a^i(t,x_0)\phi(t,x_0)\, dt \qquad (2.55)$$

since the distributional derivative of v w.r.t. x is given by

$$\partial_x v \;=\; (v_r - v_l)\delta_{x-x_0}$$

In general, we have the following weak solution of equation (2.44)

$$\gamma^i(t,x) \;=\; h(p(t,x)) \qquad (2.56)$$

where $h(x)$ denotes the initial profile, i.e., the inital data at $t = 0$. Furthermore, $p(t,x)$ corresponds to the point $(t = 0, x = \bar{x})$ such that the characteristic emanating at $(0,\bar{x})$ leads through the point (t,x) (this is only possible because of the uniqueness of the Filippov characteristic). More explicitly, we have for $s = 0$

$$p(t,x) \;=\; \begin{cases} x - v_l \cdot t & x < x_0 \\ \frac{v_l}{v_r}(x - x_0) + x_0 - v_l \cdot t & x_0 < x < x_0 + v_r \cdot t \\ x - v_r \cdot t & x > x_0 + v_r \cdot t \end{cases} \qquad (2.57)$$

We discuss the possible cases for the characteristics with $s \neq 0$ in appendix A. In particular we will see that in our application we can always guarantee a unique Filippov characteristic given by (2.57). The unique weak solution to the commodity equation (2.44) is then given by transporting the initial value along the generalized characteristics as in (2.56).

In the literature there are cases in which the Filippov characteristics are not unique, cf. [93]. The simplest example for such a situation is given by considering

$$v(t,x) \;=\; +sign(x) \qquad (2.58)$$

As pointed out in [93], the Filippov characteristics are given by

$$X(t;0,\bar{x}) \;=\; \begin{cases} \bar{x} - t & \bar{x} < 0 \\ \bar{x} + t & \bar{x} > 0 \\ \pm(t - t^*)H(t - t^*) & \bar{x} = 0 \end{cases}$$

Herein $H(t)$ denotes the Heaviside function and $t^* > 0$ is arbitrary. The Filippov characteristics for this example are depicted in figure 2.10.

We can in general not construct a weak solution with the method outlined above. Transporting the initial value $\gamma(0,0)$ into the ?–region along the Filippov characteristics will not conserve mass as required. As we will see in appendix A a situation as in figure 2.10 can not occur for our particular application.

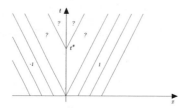

Figure 2.10: Example for non–unique Filippov–characteristics. Initial values can not simply be transported along the characteristics emanating at $\bar{x} = 0$. Conservation of mass can not be guaranteed by this procedure.

Measure solutions

To complete the discussion, we note that in general we will only obtain measure-valued solutions to the conservative equation (2.59) which is obtained by formally differentiating (2.44) w.r.t. x (provided $\gamma^i \in BV$), cf. [93]:

$$\partial_t \mu + \partial_x (v \cdot \mu) = 0 \qquad (2.59)$$

where $\mu = \partial_x \gamma^i \in \mathcal{M}$ and $\mathcal{M}$ is the set of bounded Borel measures. Note that one has to take care how to define the product $v \cdot \mu$; for details see [14] and [93].

As pointed out in [93] for the example depicted in figure 2.10 we can in general not construct a measure solution by tracing back the characteristics. This procedure is depicted in figures 2.11 and 2.12 and explained below for a simpler example. We take $t^* < 1$ in figure 2.10. Assume at $t = 0$ we have an initial measure μ_0. Consider at time $t = 1$ the two disjoint intervals $I := [a, b] = [-2, -\frac{1}{3}]$ and $J := [\bar{a}, \bar{b}] = [\frac{1}{3}, 2]$. Tracing back the characteristics, we obtain at $t = 0$ two "father" intervals $I^* := [a^*, b^*] = [-1, 0]$ and $J^* := [0, 1]$. Consequently, we have

$$\mu(1)(I) + \mu(1)(J) = \mu_0(I^*) + \mu_0(J^*) = \mu_0([-1, 0]) + \mu_0([0, 1]) \qquad (2.60)$$

On the other hand measures need to be additive for disjoint intervals. Therefore, we have

$$\mu(1)(I) + \mu(1)(J) = \mu(1)(I \cup J) = \mu_0([-1, 1]) \qquad (2.61)$$

In (2.60) we count the initial mass at the origin twice whereas in (2.61) it is accounted for only once. Consequently, the construction will –in this example– conserve mass solely if $\mu_0 \in L^1_{loc}(\mathbb{R})$ or more generally

$$\mu_0(\{0\}) = 0 \qquad (2.62)$$

As stated in [93] the unique *reversible* solution for the nonconservative equation defined in [14] would be constant in the ?-region in figure 2.10. Therefore, the corresponding measure solution will vanish in this region, i.e., satisfy (2.62). Hence the total mass will be conserved.

According to a result in [93] (Definition 3.2 and Theorem 3.2), there exists a unique mass-conserving measure solution to equation (2.59) provided v satisfies the OSLC (which ensures uniqueness of the Filippov characteristics in our case). We consider (2.59) and a discontinuous velocity v as in (2.48) with $s = 0$. Additionally, we supply equation (2.59) with initial data $\mu(0, I) = \mu_0(I)$ for intervals $I = [a, b]$. The solution $\mu(\tau, [a, b])$ is obtained graphically using the uniquely defined Filippov characteristics. We follow the characteristic leading through a at time $t = \tau$ back to time $t = 0$ and find a point a^*. Analogoulsy, we follow the characteristic through b at time $t = \tau$ to find a point b^* at time $t = 0$. Then we set

$$\mu(\tau, [a, b]) \quad := \quad \mu(0, [a^*, b^*])$$

This procedure is depcited in figures 2.11 and 2.12.

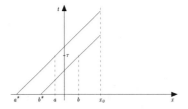

Figure 2.11: Determination of $[a^*, b^*]$: first example.

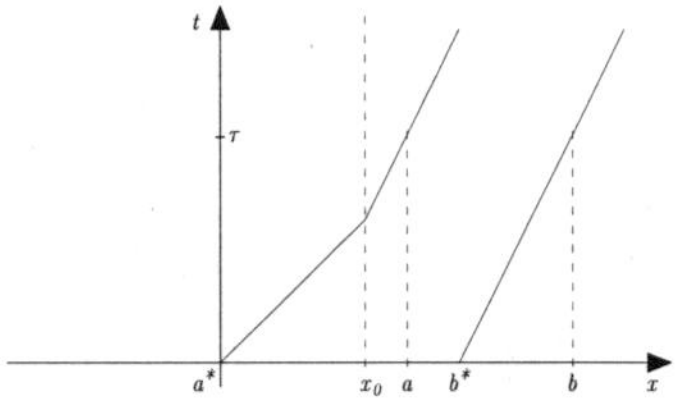

Figure 2.12: Determination of $[a^*, b^*]$: second example.

Following this idea more mathematically, we obtain for $s = 0$ in (2.48) the following measure solution to (2.59)

$$a < b < x_0 \qquad \mu(t, [a, b[) = \mu_0([a - v_l \cdot t, b - v_l \cdot t[)$$

$$a < x_0 < b < x_0 + v_r \cdot t \qquad \mu(t, [a, b[) = \mu_0([a - v_l \cdot t, x_0 - v_l \cdot t[)$$
$$+ \mu_0([x_0 - v_l \cdot t, \frac{v_l}{v_r}(b - x_0) + x_0 - v_l \cdot t[)$$

$$a < x_0 < x_0 + v_r \cdot t < b \qquad \mu(t, [a, b[) = \mu_0([a - v_l \cdot t, x_0 - v_l \cdot t)[)$$
$$+ \quad \mu_0([x_0 - v_l \cdot t, x_0[) + \mu_0([x_0, b - v_r \cdot t)[)$$

$$x_0 < a < b < x_0 + v_r \cdot t \qquad \mu(t, [a, b[) =$$
$$\mu_0([\frac{v_l}{v_r}(a - x_0) + x_0 - v_l \cdot t, \frac{v_l}{v_r}(b - x_0) + x_0 - v_l \cdot t[)$$

$$x_0 < a < x_0 + v_r \cdot t < b \qquad \mu(t, [a, b[) = \mu_0([\frac{v_l}{v_r}(a - x_0) + x_0 - v_l \cdot t, x_0[)$$
$$+ \mu_0([x_0, b - v_r \cdot t)[)$$

$$x_0 < x_0 + v_r \cdot t < a < b \qquad \mu(t, [a, b[) = \mu_0([a - v_r \cdot t, b - v_r \cdot t[)$$

Note that the **major** contribution is the scaling introduced whenever the measure is taken in the region where the characteristics have kinks. In particular, for the case $a < x_0 < x_0 + v_r \cdot t < b$ the interesting interval is $[x_0, x_0 + v_r \cdot t]$. Its length obviously is $v_r \cdot t$, the "inital" interval, i.e., the interval $[a, b]$ where $u_0([a, b])$ needs to be taken, has length $v_l \cdot t$. Accordingly we see, that the scaling factor in order to transform the too small length $v_r \cdot t$ to $v_l \cdot t$ is $\frac{v_l}{v_r}$. However, since we take μ_0 in the above formula, this factor does not appear in this case, since we have the correct length, namely $v_l \cdot t$, already.

34

If μ_0 is the Lebesgue-measure, we obviously have

$$a < b < x_0 \qquad \mu(t, [a, b[) = b - a$$

$$a < x_0 < b < x_0 + v_r \cdot t \qquad \mu(t, [a, b[) = (x_0 - a) + \frac{v_l}{v_r}(b - x_0)$$

$$a < x_0 < x_0 + v_r \cdot t < b \qquad \mu(t, [a, b[) = (x_0 - a) + v_l \cdot t + (b - x_0 - v_r \cdot t)$$

$$x_0 < a < b < x_0 + v_r \cdot t \qquad \mu(t, [a, b[) = \frac{v_l}{v_r}(b - a)$$

$$x_0 < a < x_0 + v_r \cdot t < b \qquad \mu(t, [a, b[) = \frac{v_l}{v_r}(x_0 + v_r \cdot t - a) + (b - v_r \cdot t - x_0)$$

$$x_0 < x_0 + v_r \cdot t < a < b \qquad \mu(t, [a, b[) = b - a$$

Example 2.1.16. *We take $v_l = 2$ and $v_r = 1$ and let μ_0 be induced by a function u_0. Assume we have*

$$u_0(x) = \begin{cases} 1 & x_0 - 1 < x < x_0 + 1 \\ 0 & else \end{cases} \tag{2.63}$$

We recall that the corresponding measure μ_0 is given by

$$\mu_0([a, b[) = \int_a^b u_0(x)\, dx \tag{2.64}$$

Hence μ_0 gives for $(a, b) \subset [x_0 - 1, x_0 + 1]$ the length $b - a$. We inspect the measure solution at $t = 1$. Following the characteristics and using the formulas from above we have

$$\begin{aligned}
\mu(1, \mathbb{R}) &= \mu(1, [x_0 + \frac{1}{2}, x_0 + 2]) \\
&= \mu(1, [x_0 + \frac{1}{2}, x_0 + 1[) + \mu(1, [x_0 + 1, x_0 + 2]) \\
&= \frac{2}{1} \cdot \frac{1}{2} + 1 \tag{2.65} \\
&= 2 \tag{2.66}
\end{aligned}$$

and thus the mass is conserved. Note that without the factor $\frac{2}{1} = \frac{v_l}{v_r}$ the mass would not be conserved and we would end up with a wrong *solution.*

Remark 2.1.17. *As already mentioned, we have the correspondence $\partial_x \gamma^i = \mu$ between (2.44) and (2.59). Thus, in principle it should be possible to construct the solution γ^i from the measure solution. However, the equality can not be understood pointwise, i.e., to hold for $x \in \mathbb{R}$. It is an equality in the space $\mathcal{M}$ of bounded Borel measures. Since $\mathcal{M}$ is the dual of $C(\mathbb{R})$, the equality reads for a fixed $t \in [0, T]$ (and arbitrary interval $I = [a, b] \subset \mathbb{R}$)*

$$\int_I \psi(x)\, d(\partial_x \gamma^i(t, x)) = \langle \mu, \psi \rangle = \int_I \psi(x)\, d\mu(t, x) \qquad \forall\, \psi \in C(\mathbb{R}) \tag{2.67}$$

More details on the solution of the commodity equations in our particular setup as well as information on reversible and duality solutions can be found in appendix A.

2.1.4 The special case of two classes

We consider –as in the preceding subsections– a *reformulation* of the source–destination model proposed in [34], which then can also be seen as a particular case of the multi–class model introduced and discussed for example in [108, 105], see below for the details. The reformulation of the multi–class or source–destination model of [34] allows for an alternative extension to road networks. In particular, the discussion of suitable coupling conditions for this model at road intersections has to be adapted to the reformulation and we introduce a different modeling of the coupling conditions, cf. Section 2.1.4. Similar to the approach in [62, 54, 19], existence of solutions is granted by an analysis of the arising (half–) Riemann problems. Moreover, we identify supply and demand functions [81, 26, 27] in this context to formulate suitable optimization problems used in the construction of the solution to the Riemann–problem. For numerical results we refer the reader to subsection 2.3.1.

Preliminary Discussion

We derive the source–destination model of [34] by starting from the multi–class model of [108, 105]. The relation between both models is given below in remark 2.1.18.

A multi–class (or multi–population) model based on the LWR–equation [85] for different car classes has been introduced in [108, 105, 10]. They consider a continuum model for traffic flow with heterogeneous m media. Therein, ρ_j is the density of the jth class and the velocity field v_j of the jth class is assumed to be a function of all densities $\{\rho_j\}_{j=1}^m$. Then, the mass conservation for the jth class yields the multi–class LWR–type model:

$$\partial_t \rho_j + \partial_x(\rho_j v_j) = 0, \; j = 1, \ldots, m. \tag{2.68}$$

The velocity fields v_j are required to satisfy additional conditions to obtain a hyperbolic system and we refer the reader to [108] and [10] for more details. A particular choice for the velocity fields is given by

$$v_j((\rho_j)_{j=1}^m) = v_j(\sum_{j=1}^m \rho_j) = v_j^f \cdot \left(1 - \frac{\sum_j \rho_j}{\rho_{max}}\right), \; j = 1, \ldots, m, \tag{2.69}$$

where v_j^f is the constant class dependent free flow velocity and ρ_{max} is the maximal density for the total (car-)density

$$\rho := \sum_j \rho_j. \tag{2.70}$$

In this case and for $m = 1$ equations (2.68) and (2.69) obviously reduce to the standard LWR traffic flow model:

$$\partial_t \rho + \partial_x(\rho v(\rho)) = 0. \tag{2.71}$$

Note that if v_j^f is the same for all cars, i.e., car species have the same fundamental diagram $(\rho_j, \rho_j v(\rho))$, then, the model (2.68,2.69) can be interpreted as a model for a car species j having properties which are **independent** of the dynamics of the actual class j. For instance, one can think of classes of cars having different destinations d_j. A model for classes of cars with different source–destination pairings has been introduced recently, cf. [34]. We recall this model in Remark 2.1.18 and show its relation to the multi–class model stated above.

Remark 2.1.18. *In [34] a source–destination model for road networks is considered. They introduce functions $\pi^j, j = 1, \ldots, m$ which specify the amount of car density ρ going from a specific source s_j to a destination d_j. The dynamics for the car density is governed by the LWR–equation and, therefore, the following system for a single road is derived:*

$$\partial_t \rho + \partial_x(\rho v(\rho)) = 0, \tag{2.72a}$$

$$\partial_t \pi^j + v(\rho)\partial_x \pi^j = 0, \ j = 1, \ldots, m, \tag{2.72b}$$

$$\sum_j \pi^j = 1. \tag{2.72c}$$

If we define

$$\rho_j := \pi^j \rho, \tag{2.73}$$

then at least formally,

$$\partial_t \rho_j = \pi^j \partial_t \rho + \rho \partial_t \pi^j = -\partial_x(\rho_j v(\rho)), \tag{2.74}$$

i.e., we recover the multi–class model (2.68) for the particular velocity fields

$$v_j \equiv v, \ j = 1, \ldots, m. \tag{2.75}$$

In [34] the extension of (2.72) to networks is performed by defining coupling conditions at road intersections. In this text, and in contrast to the work of [34], we discuss the extension of the reformulation (2.68),(2.75) to networks and present numerical results. This allows for an alternative view on the coupling conditions and induces in particular the possibility to define class–specific distribution rates of cars at junctions.

Having the previous remark in mind, we consider the reformulated version of the source–destination (resp. multi–class) model. In this study, we restrict ourselves to the case $m = 2$. We normalize the total density $\rho_{max} = 1$ and set the free flow velocities $v_1^f = v_2^f = c$, $c > 0$ constant. Hence, we might allow different total free flow velocities on different parts of the road network. Finally, we obtain the following system for the evolution of a two–class model on a single road $x \in [a, b], t > 0$,

$$\frac{\partial}{\partial t}\begin{pmatrix} \rho_1 \\ \rho_2 \end{pmatrix} + \frac{\partial}{\partial x}\begin{pmatrix} \rho_1 v(\rho) \\ \rho_2 v(\rho) \end{pmatrix} = \begin{pmatrix} 0 \\ 0 \end{pmatrix}, \tag{2.76}$$

where $\rho = \rho_1 + \rho_2$ and (cf. (2.69)),

$$v(\rho) = v(\rho_1 + \rho_2) = c\left(1 - (\rho_1 + \rho_2)\right). \tag{2.77}$$

For the discussion of suitable coupling conditions at a road intersection, we need to recall the following properties of the hyperbolic system (2.76). Let $U = (\rho_1, \rho_2)$. The eigenvalues are readily computed, cf. [108]. We obtain

$$\lambda_1(U) = c \cdot (1 - 2(\rho_1 + \rho_2)) \leq \lambda_2(U) = c \cdot (1 - (\rho_1 + \rho_2)). \tag{2.78}$$

Hence, away from the vacuum $U = (0,0)$, the system is strictly hyperbolic. The eigenvectors corresponding to λ_1 and λ_2 are, respectively

$$r_1(U) = \begin{pmatrix} \frac{\rho_1}{\rho_2} \\ 1 \end{pmatrix}, \qquad r_2(U) = \begin{pmatrix} -1 \\ 1 \end{pmatrix} \tag{2.79}$$

and the characteristic family associated with the first eigenvalue is genuinely nonlinear (**GNL**) whereas the second characteristic family is linearly degenerate (**LD**). The Riemann invariants in the sense of Lax (RI-Lax) are

$$w(U) = \log\left(\frac{\rho_1}{\rho_2}\right) \quad \text{and} \quad z(U) = c \cdot (1 - (\rho_1 + \rho_2)) = v(U), \tag{2.80}$$

respectively. Next, we state the elementary waves [24] of (2.76).

An arbitrary left state $U^- = (\rho_1^-, \rho_2^-) > 0$ can be connected to a state U^+ by a 1-rarefaction wave, iff $U^+ = U(\xi)$ for $0 < \xi \leq \rho_1^-$ and where $\xi \to U(\xi)$ is the following parametrization of the 1-rarefaction wave curve:

$$U(\xi) = \begin{pmatrix} 1 \\ \frac{\rho_2^-}{\rho_1^-} \end{pmatrix} \xi, \qquad 0 < \xi \leq \rho_1^-. \tag{2.81}$$

Therefore, in the phase plane (ρ_1, ρ_2), the 1-rarefaction curve is a straight line of slope $\frac{\rho_2^-}{\rho_1^-}$ as depicted in Figure 2.13.

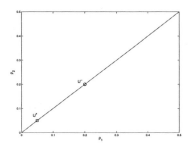

Figure 2.13: The 1-rarefaction and 1-shock curves in the phase plane (ρ_1, ρ_2).

Next, we consider the 1-(Lax)-shock waves: A left state $U^- = (\rho_1^-, \rho_2^-) > 0$ can be connected to any other state $U^+ = (\rho_1^+, \rho_2^+) > 0$ by a 1-shock wave, iff $U^+ = U(\xi)$ for some ξ where $\xi \to U(\xi) = (\rho_1(\xi), \rho_2(\xi))$ is a parametrization of the 1-shock wave curve:

$$U(\xi) = \begin{pmatrix} 1 \\ \frac{\rho_2^-}{\rho_1^-} \end{pmatrix} \xi, \qquad \rho_1^- \leq \xi \leq \frac{\rho_1^-}{\rho_1^- + \rho_2^-}. \tag{2.82}$$

The shock speed s_1 is given by

$$s_1 = c \left(1 - (\rho_1(\xi) + \rho_2(\xi) + \rho^-)\right), \tag{2.83}$$

and for system (2.76), the 1–rarefaction and 1–shock wave curves coincide, cf. Figure 2.13; therefore (2.76) is a Temple system [98].

It remains to discuss the 2-contact discontinuities. A left state $U^- = (\rho_1^-, \rho_2^-)$ can be connected to $U^+ = (\rho_1^+, \rho_2^+)$, iff $U^+ = U(\xi)$ where $U(\xi)$ is given by

$$U(\xi) \;=\; \begin{pmatrix} 1 \\ -1 \end{pmatrix} \cdot \xi + \begin{pmatrix} 0 \\ \rho^- \end{pmatrix} \qquad 0 \le \xi \le \rho^-. \tag{2.84}$$

Of course, $v(U^+) = v(U(\xi)) = v(U^-)$ for all $0 \le \xi \le \rho^-$ and in the phase plane the 2-contact wave curve is a straight line through U^- with slope -1.

Now, a Riemann problem for (2.76) is a Cauchy problem for (2.76) with the piecewise constant initial data

$$U_0(x) = \begin{pmatrix} U^- & x < 0 \\ U^+ & x > 0 \end{pmatrix}. \tag{2.85}$$

From the above discussion, we obtain that the Riemann problem with data U^- and U^+ and such that $0 < \rho^- \le 1$ and $0 < \rho^+ \le 1$, always admits a weak entropy solution [77]. In general, a solution is a composition of a 1-(Lax-)shock or 1–rarefaction wave connecting U^- to U^*,

$$U^* := \frac{\rho^+}{\rho^-} \begin{pmatrix} \rho_1^- \\ \rho_2^- \end{pmatrix}, \tag{2.86}$$

and a 2–contact discontinuity connecting U^* with U^+, see Figure 2.14. Hence, we call initial data $U = (\rho_1, \rho_2)$ **admissible**, if it satisfies $0 < \rho_1 + \rho_2 \le 1$.

Proposition 2.1.19. *Given admissible initial data U^- and U^+, the Riemann problem for (2.76) admits a unique weak entropy solution $U(x,t)$, cf. figure 2.14.*

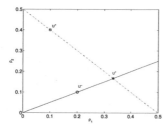 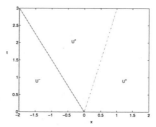

Figure 2.14: The solution to a Riemann problem in the phase plane (left) and in the (x, t) plane (right).

Before we discuss the extensions of the above model to a network, we introduce the notion of supply and demand functions, cf. [81, 26]. This will be used later to determine the range of admissible states at a junction, see below. The supply and demand functions are defined in the $(\rho, \rho v)$-plane with $\rho = \rho_1 + \rho_2$ and ρv being the total flux. Recalling Remark 2.1.18 (and [34], respectively), it is not surprising, that the 1– and 2–wave curves allow for a particular simple characterization in the $(\rho, \rho v)$-plane.

Proposition 2.1.20. *Given an admissible left state $U^- = (\rho_1^-, \rho_2^-)$, there is a one–to–one correspondence between the states U^+ which can be connected by either a 1–rarefaction or 1–shock wave to U^- and the points $(\eta, \eta v(\eta))$, $0 < \eta \leq 1$, (with $\eta = \rho(\xi)$) of the $(\rho, \rho v)$–plane. Here, $v(\eta) = c(1 - \eta)$ is as in (2.77).*

Proof. A state U^+ can be connected by a 1–rarefaction or 1–shock wave to U^-, iff $U^+ = U(\xi)$ and $U(\xi)$ is given by either (2.81) or (2.82), i.e.,

$$U(\xi) = \begin{pmatrix} 1 \\ \rho_2^-/\rho_1^- \end{pmatrix} \xi, \qquad 0 < \xi \leq \frac{\rho_1^-}{\rho_1^- + \rho_2^-}. \tag{2.87}$$

Therefore, $\rho(\xi) = \rho_1(\xi) + \rho_2(\xi) = \xi(1 + \rho_2^-/\rho_1^-)$ and satisfies $0 < \eta := \rho(\xi) \leq 1$. $\qquad\square$

The mapping $\eta \to \eta v(\eta)$ is usually referred to as fundamental diagram. Similarly, we obtain the following assertion for the states U^+ which can be connected to U^- by a wave of the second family:

Proposition 2.1.21. *Given an admissible left state $U^- = (\rho_1^-, \rho_2^-)$, the set of all states U^+ which can be connected to U^- by a wave of the second family is mapped to the single point $(\eta, \eta\, v(\eta))$, $\eta = \rho_1^- + \rho_2^-$, in the $(\rho, \rho v)$–plane.*

Proof. This is immediate since $U^+ = U(\xi)$ with $U(\xi)$ given by equation (2.84). Hence the state $U^+ = U(\xi) = (\xi, -\xi + \rho^-)$ satisfies $\rho(\xi) = \rho^-$. $\qquad\square$

As in [54, 81], we introduce the demand and supply functions defined in the $(\rho, \rho v)$–plane:

Definition 2.1.22. Supply and Demand function *The demand function $\eta \to d(\eta)$ and the supply function $\eta \to s(\eta)$ are the non-decreasing (see Figure 2.15) and the non-increasing (see Figure 2.16) parts of the curve $\eta \to \eta v(\eta)$ in the $(\rho, \rho v)$-plane, respectively.*

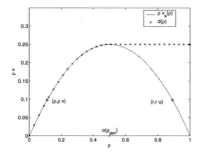
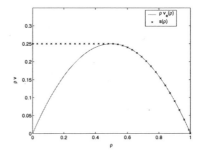

Figure 2.15: Demand function in the $(\rho, \rho v)$ plane.

Figure 2.16: Supply function in the $(\rho, \rho v)$ plane.

This finishes the characterization of states $U^\pm$ on the 1– and 2–wave curves. We will reconsider the supply and demand notion when defining solutions to a network problem.

40

Extension to a network

As in [62] and subsection 2.1.1, we model a road network as a finite direct graph $\mathcal{G} = (V, A)$ with $|V| = \mathbf{I}$ and $|A| = \mathbf{N}$. Each arc $j \in A$ corresponds to a road and each vertex $v \in V$ to a junction. For a fixed junction $v \in V$, we denote by δ_v^- the set of all its incoming roads whereas δ_v^+ is the set of all its outgoing roads. We parameterize each road $j \in A$ by an interval $[a_j, b_j]$ with possibly $a_j = -\infty$ or $b_j = \infty$. At a fixed vertex v we set $x_v^k = b_k$ if $k \in \delta_v^-$ and $x_v^k = a_k$, if $k \in \delta_v^+$.

Contrary to the notation in section 2.1.1 we use upper indices to indicate the road–dependence throughout this section. On each road $j \in A$ we consider system (2.88)

$$\partial_t \begin{pmatrix} \rho_1^j \\ \rho_2^j \end{pmatrix} + \partial_x \left(\begin{pmatrix} \rho_1^j \\ \rho_2^j \end{pmatrix} v^j(\rho^j) \right) \;=\; 0, \quad x \in [a_j, b_j], \ t > 0, \tag{2.88}$$

where as before

$$U^j := (\rho_1^j, \rho_2^j), \quad \rho^j = \rho_1^j + \rho_2^j, v^j(\rho^j) = c^j(1 - \rho^j), \tag{2.89}$$

and for notational convenience $F^j(U) = (\rho_1, \rho_2)v^j(U)$. Note, that we explicitly allow for different velocity profiles on different roads (modeled by arc dependent constants c^j).

As in [62] and analogous to definition 2.1.3 we call a set of functions $\{U^j\}_{j \in A}$ a weak solution of system (2.88), if it satisfies

$$\sum_{j \in A} \int_0^\infty \int_{a_j}^{b_j} U^j \cdot \partial_t \phi^j + F^j(U^j) \cdot \partial_x \phi^j \, dx \, dt \;=\; 0, \tag{2.90}$$

for all families of test functions $\{\phi^j\}_{j \in A}$ where each function $\phi^j : \mathbb{R}^+ \times [a_j, b_j] \to \mathbb{R}^2$ is compactly supported in $(0, \infty) \times [a_j, b_j]$ and is smooth across a vertex $v \in V$: $\phi^k(x_v^k, t) = \phi^i(x_v^i, t)$ for $k \in \delta_v^-$ and $i \in \delta_v^+$.

From (2.90) we derive the Rankine–Hugoniot conditions (2.91) as coupling conditions at a vertex $v \in V$:

$$\sum_{k \in \delta_v^-} \rho_i^k v^k(\rho^k)(x_v^k, t) \;=\; \sum_{l \in \delta_v^+} \rho_i^l v^l(\rho^l)(x_v^l, t), \ t > 0, \ i = 1, 2. \tag{2.91}$$

The conditions (2.91) state the conservation of each class of cars through the intersection. Frequently, we will call $\rho_i^k v^k(\rho^k)$, the i-th moment. Depending on the degree of the vertex, additional conditions have to be imposed to obtain a unique solution.

A major step in the construction of a network solution is the consideration of half-Riemann problems at a junction. A half-Riemann problem at a vertex $v \in V$ is obtained by considering $|\delta_v^-| + |\delta_v^+|$ Riemann problems, one on each arc $j \in \delta_v^- \cup \delta_v^+$, and each arc considered as extended to $(-\infty, \infty)$. Depending whether the arc is in– or outgoing to the vertex, either the left or the right data is given by the initial data $U_{k,0}$, i.e., we consider the Riemann problems:

$$\frac{\partial}{\partial t} U^k + \frac{\partial}{\partial x} F^k(U^k) \;=\; 0, \tag{2.92a}$$

$$U^k(x, 0) \;=\; \begin{cases} U^- & x < x_v^k, \\ U^+ & x > x_v^k, \end{cases} \tag{2.92b}$$

where for $k \in \delta_v^-$, **only** the data U^- is given by the initial data, and for $k \in \delta_v^+$ **only** the data U^+ is given by the initial data $U_{k,0} = U^k(x, 0)$.

Below, in proposition 2.1.23 and 2.1.24 we identify possible states U^+ and U^-, respectively, such that the waves in the solution to (2.92) have either non–positive ($k \in \delta_v^-$) or non–negative speed ($k \in \delta_v^+$). It turns out that the simplest way to describe the admissible states $U^\pm$ employs the notion of supply and demand function, cf. definition 2.1.22. In the sequel, we drop the index k whenever the intention is clear.

Admissible Riemann data for in- and outgoing roads

First, we consider for a $v \in V$ the case $k \in \delta_v^-$, i.e., we solve the Riemann problems for an ingoing road to a vertex $v \in V$. Here the initial data U^- for (2.92) are given ($U^- = U_{k,0}$) and we determine all states U^+, such that the solution $U(x, t)$ to (2.92) is either a constant or contains waves of negative speed only. We call such states U^+ **admissible for the incoming road**. For the same reasoning as in [62] we excluded stationary shocks. Since the wave speed of waves of the second family is $v(U) \geq 0$, the only states U^+ which can be connected to U^- have to be on the 1–wave curve through U^-. More precisely, we have:

Proposition 2.1.23. *Let $U^- = (\rho_1^-, \rho_2^-)$ be an admissible initial value (i.e., $U^- \neq 0$, $\rho_1^- + \rho_2^- \leq 1$) on an incoming road $k \in \delta_v^-$. Let the demand function d be given as in definition 2.1.22. Then, for any given flux $q \in \mathbb{R}$, such that*

$$0 < q \leq d(\rho^-), \tag{2.93}$$

there exists exactly one state U^+ which is admissible for the incoming road and satisfies $\rho^+ v(\rho^+) = q$.

Proof. Since we neglect stationary shocks, we set $U^+ \equiv U^-$ in the case $q = d(\rho^-)$. Assume now $\eta < d(\rho^-)$. Due to Proposition 2.1.20 there is a correspondence between the curves in the $(\rho_1, \rho_2)-$ plane and the $(\rho, \rho v)-$plane. It remains to prove that this is also true for wave speeds of the 1–waves: If $U^+ = (\rho_1^+, \rho_2^+)$ is connected to U^- by a 1–rarefaction wave, the corresponding wave speed is given by

$$\lambda_1(U) = c(1 - 2(\rho_1 + \rho_2)) = \partial_\eta \left(\eta v(\eta) \right)_{\eta = \rho_1 + \rho_2}. \tag{2.94}$$

If U^+ is connected to U^- by a 1–shock, the corresponding shock speed is given by equation (2.83), i.e.,

$$s_1 = c(1 - \rho_1^+ - \rho_2^+ - \rho_1^- - \rho_2^-) \ = \ \frac{(\eta v(\eta))_{\eta = \rho^+} - (\eta v(\eta))_{\eta = \rho^-}}{\rho^+ - \rho^-} = \frac{[\eta v(\eta)]}{[\eta]}. \tag{2.95}$$

Therefore, the wave speeds of the solution to a Riemann problem with initial data U^- and U^+ can also be obtained from the fundamental diagram in the $(\rho, \rho v)-$plane. Hence, we can equivalently discuss a Riemann problem for the first–order LWR model $\partial_t \eta + \partial_x (\eta v(\eta)) = 0$ with left Riemann data given by η^- for $\eta^- = \rho_1^- + \rho_2^-$. Then, the assertion of the proposition is immediate, see e.g. [62, 54]. $\square$

Similarly, we obtain a result for admissible states for an outgoing road $k \in \delta_v^+$ at a node $v \in V$. In this case the initial data U^+ for (2.92) is given $(U^+ = U_{k,0})$ and we determine all states U^-, such that the solution $U(x,t)$ to (2.92) is either a constant or contains waves of positive speed only. Again, we call such states U^- **admissible for the outgoing road** and exclude stationary shocks and the vacuum from the discussion.

Proposition 2.1.24. *Consider admissible initial data $U^+ = (\rho_1^+, \rho_2^+)$ and an outgoing road $k \in \delta_v^+$ at a node $v \in V$. Let the supply function s be given as in definition 2.1.22. For an arbitray given flux $q \in \mathbb{R}$ we assume*

$$0 < q \le s(\rho^+). \tag{2.96}$$

Then a state $U^- = (\rho_1^-, \rho_2^-)$ is admissible for the outgoing road, iff $\rho_1^- + \rho_2^- = \eta^-$. Herein, η^- is the unique value $\eta^- \in (0, 1]$, such that $q = \eta^- v(\eta^-)$ and such that the solution to the Riemann problem for the scalar LWR model

$$\partial_t \eta + \partial_x (\eta v(\eta)) = 0, \eta_0(x) = \begin{pmatrix} \eta^- & x < 0 \\ \rho^+ & x > 0 \end{pmatrix}, \tag{2.97}$$

consists of waves of non–negative speed.

Proof. The proof is slightly more involved than in the previous case, since a solution $U(x,t)$ to (2.92) is in general a composition of waves of the first *and* second family. First note, that as in [62, 54], we obtain a unique value η^- with the properties stated above. Now, assume we have a Riemann datum $U^- = (\rho_1^-, \rho_2^-)$ which satisfies $\eta^- = \rho_1^- + \rho_2^-$. As in the previous paragraph the Riemann problem (2.92) admits a solution $U(x,t)$ with possible intermediate state $U^* = \rho^+ / \rho^- \begin{pmatrix} \rho_1^- \\ \rho_2^- \end{pmatrix}$. Obviously, the 2–wave connecting U^* and U^+ has non–negative speed. Moreover, as in the proof of the previous proposition, the speed of the *1–wave* connecting U^- to U^* is given by the wave speed of the solution to a Riemann problem for the scalar LWR–equation (2.97) with initial data (ρ^-, ρ^*). Now, $\rho^- = \eta^-$ and due to proposition 2.1.21, $\rho^* = \rho^+$. Hence, we have $s(\rho^*) = s(\rho^+)$ and by construction of η^- the solution to (2.97) consists of waves of non–negative speed. $\square$

In proposition 2.1.24 we can not guarantee the uniqueness for the admissible states $U^- = (\rho_1^-, \rho_2^-)$, only its total density $\rho_1^- + \rho_2^-$ is uniquely given by η^-. Again, propositions 2.1.23 and 2.1.24 show the fact, that the proposed model is a reformulation of the LWR–model combined with an additional advection equation. However, in the case of road intersections the proposed reformulation proves useful, cf. the following discussion.

The 1–1 junction

We consider a network consisting of just one junction with degree 2 ($|\delta_v^-| = 1$ and $|\delta_v^+| = 1$) and construct a weak solution in the sense of (2.90) for constant initial data $U_0^j, j = 1, 2$. Note, that the different arcs $j = 1, 2$ might have different free flow velocities c^j, cf. (2.89), due to changed road conditions. We recall that we have by definition $x_v^1 = x_v^2$ and $\rho^j = \rho_1^j + \rho_2^j$.

Proposition 2.1.25. *Let U_0^j, $j = 1, 2$ be some given admissible initial data, constant on each arc $j = 1, 2$. Then there exists a unique weak solution $\{U^1 = (\rho_1^1, \rho_2^1), U^2 = (\rho_1^2, \rho_2^2)\}$ in the sense of (2.90) with the following properties:*

1. *Both moments are conserved through the junction, i.e., (2.91) holds for $t > 0$.*

2. *The total flux $\rho^1 v^1(\rho^1) = \rho^2 v^2(\rho^2)$ is maximal at the interface $(x, t) = (x_v^1, t), t > 0$.*

Proof. We define $U^{1-} := U_0^1$ and $U^{2+} := U_0^2$ and consider the following maximization problem for the unknowns $\rho_1^{1+}, \rho_2^{1+}, \rho_1^{2-}$ and ρ_2^{2-}:

$$\max \rho^{1+} v^1(\rho^{1+}) \text{ subject to} \tag{2.98a}$$
$$\rho_1^{1+} v^1(\rho^{1+}) = \rho_1^{2-} v^2(\rho^{2-}), \tag{2.98b}$$
$$\rho_2^{1+} v^1(\rho^{1+}) = \rho_2^{2-} v^2(\rho^{2-}), \tag{2.98c}$$
$$0 < \rho^{1+} v^1(\rho^{1+}) \leq d(\rho^{1-}), \tag{2.98d}$$
$$0 < \rho^{2-} v^2(\rho^{2-}) \leq s(\rho^{2+}). \tag{2.98e}$$

Here, s and d denote the supply and demand function for the outgoing and incoming road, respectively, cf. definition 2.1.22. Introducing $q^1 := \rho^{1+} v^1(\rho^{1+})$ we obtain $q^1 = \min\{d(\rho^{1-}), s(\rho^{2+})\}$ to be the maximal total flux at the interface. Due to Proposition 2.1.23 we obtain a unique state $U^{1+} = (\rho_1^{1+}, \rho_2^{1+})$, such that $\rho^{1+} v^1(\rho^{1+}) = q^1$. Due to proposition 2.1.24 we obtain a unique total density $(\eta^- =) \rho^{2-}$ such that $\rho^{2-} v^2(\rho^-) = q^1$. Moreover, (2.98b) and (2.98c) then uniquely determine ρ_1^{2-} and ρ_2^{2-}, since $v^2(\rho^{2-}) > 0$ by (2.98e). Finally, U^1 is obtained as solution to the half–Riemann problem (2.92) with initial data $U^- := U^{1-} \equiv U_0^1$ and $U^+ := U^{1+}$. The solution U^2 is obtained as solution to (2.92) with initial data $U^- := U^{2-}$ and $U^+ := U^{2+} \equiv U_0^2$. Due to Proposition 2.1.23 and 2.1.24, U^j is a superposition of waves of the first and (for $j = 2$) second family having only non–positive (resp. non–negative) wave speeds. We have $U^1(x_v^1, t) = U^{1+}$ and $U^2(x_v^2, t) = U^{2-}$ for $t > 0$ and therefore, by construction, the assertions of the proposition are fulfilled. $\square$

Remark 2.1.26. *In the case of $c^j = c$, $j = 1, 2$ the junction is artificial and a weak solution to (2.90) is given by the entropy solution of a standard Riemann problem for (2.76) with initial data $U_0 = \begin{pmatrix} U_0^1 & x < x_v^1 \\ U_0^2 & x > x_v^2 \end{pmatrix}$. The solution to the standard Riemann problem coincides with the one obtained by the previous proposition 2.1.25.*

The 2–1 and 1–2 junction

First, we consider a network consisting of three connected arcs $j = 1, 2, 3$ at a vertex $v \in V$ with $|\delta_v^-| = 2$, i.e., a situation where two roads merge into one other road ($j = 3$). As before, we consider the case of constant initial data $U_0^j, j = 1, 2, 3$. Proposition 2.1.27 guarantees the existence of a weak solution in the sense of (2.90). In fact, the only relevant change to the previous discussion is the formulation of a suitable maximization problem to obtain the right ($j = 1, 2$) and left ($j = 3$) initial data for the half-Riemann problems (2.92). In the current case, we maximize the total incoming flux. We do not obtain a unique solution, since the maximization does *not* necessarily possess a *unique* solution. It is possible to introduce additional conditions to obtain uniqueness, as for example below, in the case of a vertex of arbitrary degree. Such conditions are subject to the particular modeling of the road intersection and have been studied for example in [19, 51, 54].

Proposition 2.1.27. *Let U_0^j, $j = 1, 2, 3$ be some given admissible initial data, constant on each arc $j = 1, 2, 3$. Then, there exists a (not necessarily unique!) weak solution $\{U^1 = (\rho_1^1, \rho_2^1), U^2 = (\rho_1^2, \rho_2^2), U^3 = (\rho_1^3, \rho_2^3)\}$ in the sense of (2.90) with the following properties:*

1. *The moments are conserved through the junction, i.e., (2.91) holds for $t > 0$.*

2. *The total flux $\rho^1 v^1(\rho^1) + \rho^2 v^2(\rho^2)$ is maximal at the interface $(x, t) = (x_v^1, t), t > 0$.*

Proof. The proof is analogous to the proof of Proposition 2.1.25, but instead of (2.98) we consider the following problem for the unknowns U^{1+}, U^{2+}, U^{3-} and with given data $U^{1-} = U_1^0, U^{2-} = U_2^0$ and $U^{3+} = U_3^0$:

$$\max \rho^{2+} v^2(\rho^{2+}) + \rho^{1+} v^1(\rho^{1+}) \text{ subject to} \tag{2.99a}$$
$$\rho_1^{1+} v^1(\rho^{1+}) + \rho_1^{2+} v^2(\rho^{2+}) = \rho_1^{3-} v^3(\rho^{3-}), \tag{2.99b}$$
$$\rho_2^{1+} v^1(\rho^{1+}) + \rho_2^{2+} v^2(\rho^{2+}) = \rho_2^{3-} v^3(\rho^{3-}), \tag{2.99c}$$
$$0 < \rho^{1+} v^1(\rho^{1+}) \leq d_1(\rho^{1-}), \tag{2.99d}$$
$$0 < \rho^{2+} v^2(\rho^{2+}) \leq d_2(\rho^{2-}), \tag{2.99e}$$
$$0 < \rho^{3-} v^3(\rho^{3-}) \leq s_3(\rho^{3+}). \tag{2.99f}$$

Again, d_j is the demand function on roads $j = 1, 2$ and s_j is the supply function on road $j = 3$ as in definition 2.1.22. Assume that $s_3(\rho^{3+}) \geq d_1(\rho^{1-}) + d_2(\rho^{2-})$. Then, as in the previous proof, problem (2.99) has a unique solution. We refer to [51, 19, 54] for additional constraints guaranteeing uniqueness in the case $s_3(\rho^{3+}) \leq d_1(\rho^{1-}) + d_2(\rho^{2-})$. For any solution to (2.99) we proceed as before and obtain a weak solution to (2.90) as solution to the three half–Riemann problems (2.92) with initial data U^{j-} and U^{j+} for $j = 1, 2, 3$, respectively. $\qquad\square$

Remark 2.1.28. *In [34] the existence of a unique weak solution at an arbitrary junction has been proven. This is achieved by introducing additional constraints and a modified objective function in the maximization problem (2.99). We propose a different choice below for a vertex of arbitrary degree, i.e., an "equal priority rule" as in [54].*

Before discussing the general case, we consider a vertex of degree three with two outgoing roads $j = 2, 3$. For this case we propose the following modeling: We introduce a distribution rate $\alpha \in [0, 1]$ of the total incoming flux $\rho(x_v^1, t) v^1(\rho(x_v^1, t))$. To be more precise, we prove existence of a weak solution $\{U^j\}_{j=1}^3$ which satisfies at a node $v \in V$ with $|\delta_v^-| = 1$ and $|\delta_v^+| = 2$

$$\alpha\, \rho_1^1 v^1(\rho^1)(x_v^1, t) = \rho_1^2 v^2(\rho^2)(x_v^2, t), \tag{2.100a}$$
$$\alpha\, \rho_2^1 v^1(\rho^1)(x_v^1, t) = \rho_2^2 v^2(\rho^2)(x_v^2, t), \tag{2.100b}$$
$$(1 - \alpha)\, \rho_1^1 v^1(\rho^1)(x_v^1, t) = \rho_1^3 v^3(\rho^3)(x_v^3, t), \tag{2.100c}$$
$$(1 - \alpha)\, \rho_2^1 v^1(\rho^1)(x_v^1, t) = \rho_2^3 v^3(\rho^3)(x_v^3, t). \tag{2.100d}$$

Obviously, (2.100) implies (2.91) and we have a distribution as $\alpha \rho^1 v^1(\rho^1) = \rho^2 v^2(\rho^2)$. Under the additional assumption (2.100), the following result holds true.

Proposition 2.1.29. *Let U_0^j, $j = 1, 2, 3$ be some given admissible initial data, constant on each arc $j = 1, 2, 3$, and a distribution rate $0 \leq \alpha \leq 1$. Then, there exists a unique weak solution $\{U^1 = (\rho_1^1, \rho_2^1), U^2 = (\rho_1^2, \rho_2^2), U^3 = (\rho_1^3, \rho_2^3)\}$ in the sense of (2.90) with the following properties:*

1. *The moments are conserved through the junction, i.e., (2.91) holds for $t > 0$, and they are distributed according to α, i.e., (2.100) holds.*

2. *The total flux $\rho^1 v^1(\rho^1)$ is maximal at the interface $(x, t) = (x_v^1, t), t > 0$.*

The proof is similar to the one of the two previous propositions and is therefore omitted. For later reference, we remark that the maximal flux at the interface $x = x_v^i$ for any time $t > 0$ in the non-trivial case $\alpha \in (0, 1)$, is given by

$$q(t) = q = \min\{d_1(\rho^1), \frac{1}{\alpha} s_2(\rho^2), \frac{1}{1-\alpha} s_3(\rho^3)\}, \tag{2.101}$$

where $\rho^j = \rho_{1,0}^j + \rho_{2,0}^j$ is given by the initial data and d_j and s_j are the demand and supply functions from definition 2.1.22, respectively.

Remark 2.1.30. *The following extension in the spirit of section 2.1.2 to the single distribution rate α is possible, cf. (2.32): We introduce two rates α^1 and α^2, both in $[0, 1]$, and distribute the partial incoming fluxes $\rho_1^1 v^1(\rho^1)$ and $\rho_2^1 v^1(\rho^1)$ with possibly different rates α^1 and α^2, i.e., we replace the equations in (2.100) by*

$$\alpha^1 \, \rho_1^1 v^1(\rho^1)(x_v^1, t) = \rho_1^2 v^2(\rho^2)(x_v^2, t),$$
$$\alpha^2 \, \rho_1^1 v^1(\rho^1)(x_v^1, t) = \rho_2^2 v^2(\rho^2)(x_v^2, t),$$
$$(1 - \alpha^1) \, \rho_1^1 v^1(\rho^1)(x_v^1, t) = \rho_1^3 v^3(\rho^3)(x_v^3, t),$$
$$(1 - \alpha^2) \, \rho_1^1 v^1(\rho^1)(x_v^1, t) = \rho_2^3 v^3(\rho^3)(x_v^3, t).$$

With this extension the maximization problem for the interface flux q does not necessarily posess a unique solution. Uniqueness can be guaranteed under additional conditions on the states.

The general m–n junction

Combining the discussion of the 1–2 and 2–1 junction, we now state the result for a network with a single vertex $v \in V = \{v\}$ with $j \in \{1, \ldots, m\} = \delta_v^-$ incoming and $j \in \{m + 1, \ldots, n + m\} = \delta_v^+$ outgoing roads. We need to introduce additional conditions to obtain a unique solution $\{U^j\}_{j=1}^{n+m}$ in the sense of (2.90). First and as in [34], we introduce a distribution matrix $A = (\alpha_{ij})_{ij} \in Mat(\mathbb{R}^m, \mathbb{R}^n)$ such that

$$\sum_{j \in \delta_v^+} \alpha_{ij} = 1, \; \forall i \in \delta_v^-. \tag{2.102}$$

46

Second, we introduce an equal priority rule as in [54]: If cars (of either class) coming from more than one incoming road move towards the same outgoing road, we assume that they enter this road turn by turn. In particular, we model this fact by looking for weak solutions which additionally satisfy $\rho^i v^i(\rho^i) = \rho^k v^k(\rho^k)$ for all $i, k \in \delta_v^-$.

Summarizing, we have the following theorem on existence and uniqueness of a weak solution $\{U\}_{j=1}^{n+m}$ to piecewise constant initial data.

Theorem 2.1.31. *Consider admissible initial data U_0^j constant on each arc $j = 1, \ldots, n+m = \delta_v^- \cup \delta_v^+$ and a flux distribution matrix $A \in Mat(\mathbb{R}^m, \mathbb{R}^n)$ satisfying (2.102) and $\sum_i \alpha_{ij} \neq 0$.*

Then, there exists a unique weak solution $\{U^j\}$ to (2.90) which satisfies both (2.91) and the following additional properties:

1. *The moments are distributed according to A, i.e.,*

$$\sum_{i \in \delta_v^-} \alpha_{ij}\, \rho_l^i v^i(\rho^i)(x_v^i, t) = \rho_l^j v^j(\rho^j)(x_v^j, t), \qquad l = 1, 2,\ j \in \delta_v^+, t > 0. \qquad (2.103)$$

2. *An equal priority rule holds true, i.e.,*

$$\rho^i v^i(\rho^i)(x_v^i, t) = \rho^j v^j(\rho^j)(x_v^j, t),\ i, j \in \delta_v^-, t > 0. \qquad (2.104)$$

3. *The total incoming flux $\sum_{i \in \delta_v^-} \rho^i v^i(\rho^i)(x, t)$ is maximal at the interface $x = x_v^i, t > 0$.*

Thanks to (2.102) we obtain (2.91) from (2.103). The restriction $0 < \sum_i \alpha_{ij}$ has been imposed to avoid triviality due to an used outgoing road.

Proof. (*of Theorem 2.1.31*) We introduce $U^{i-} = U_0^i, i \in \delta_v^-$ and $U^{j+} = U_0^j, j \in \delta_v^+$. To prove the assertion we consider the following maximization problem in the unknowns $(U^{i+}, U^{j-})_{i,j}$ for $i \in \delta_v^-$ and $j \in \delta_v^+$.

$$\max \sum_{i \in \delta_v^-} \rho^{i+} v^i(\rho^{i+}) \text{ subject to} \qquad (2.105a)$$

$$\sum_{i \in \delta_v^-} \alpha_{ij}\, \rho_l^{i+} v^i(\rho^{i+}) = \rho_l^{j-} v^j(\rho^{j-}),\ l = 1, 2, j \in \delta_v^+, \qquad (2.105b)$$

$$0 < \rho^{i+} v^i(\rho^{i+}) \leq d_i(\rho^{i-}),\ i \in \delta_v^-, \qquad (2.105c)$$

$$0 < \rho^{j-} v^j(\rho^{j-}) \leq s_j(\rho^{j+}),\ j \in \delta_v^+, \qquad (2.105d)$$

$$\rho^{i+} v^i(\rho^{i+}) = \rho^{k+} v^k(\rho^{k+}),\ i, k \in \delta_v^-. \qquad (2.105e)$$

If we introduce $\tilde{q} := \rho^{i+} v^i(\rho^{i+}) = \rho^{k+} v^k(\rho^{k+}), i, k \in \delta_v^-$, we obtain that $\tilde{q} \in \mathbb{R}$ is uniquely determined as

$$\tilde{q} = \min\Big\{ d_i(\rho^{i-}), s_j(\rho^{j+}) / \sum_i \alpha_{ij} : i \in \delta_n^-, j \in \delta_n^+ \Big\}. \qquad (2.106)$$

Due to proposition 2.1.23 we obtain a unique state $U^{i+} = (\rho_1^{i+}, \rho_2^{i+})$ for every incoming road $i \in \delta_v^+$, such that $\rho^{i+} v^i(\rho^{i+}) = \tilde{q}$. Moreover, $U^i(x,t)$ is given as solution to the half-Riemann problem (2.92) with initial data U^{i-} and U^{i+} and consists of waves having non–positive wave speed. Then, due to (2.105b) and proposition 2.1.24 we obtain a unique state $U^{j-} = (\rho_1^{j-}, \rho_2^{j-})$ for each outgoing road $j \in \delta_v^+$ and such that the solution U^j to the half–Riemann problem (2.92) with initial data U^{j-} and U^{j+} is a superposition of waves of non–negative speed. Finally, $\{U^i\}$ fulfills (2.103) and (2.104) by construction. $\square$

2.2 Optimization Strategies

This section contains a general framework for certain optimization problems on networks. These optimization problems can be formulated on a continuous level with constraints consisting of partial or ordinary differential equations. Then, usually adjoint calculus is used for efficient computation of the optimal control. This approach has already been successfully applied in different areas. Among the vast literature we only mention some examples such as optimal control of fluid flows [60], optimal semiconductor design [61] or general initial value control of hyperbolic equations [103, 104]. For optimal control in the context of networks we refer to [49] for adjoint calculus in the context of traffic flow networks.

Efficient optimization algorithms are known in cases when one knows a descent direction. A particular descent direction can be computed with the so–called adjoint calculus. We were able to extend this known technique [78, 103, 60, 61] to networks; in fact, we developed a general framework for adjoint computations on networks. We tested this approach in the context of supply networks (cf. subsection 3.1.1 below), since the models are somewhat easier to deal with than the ones in the context of multi–class networks. The theoretical preliminaries are presented in subsection 2.2.1. Subsection 2.2.2 is to be understood as a preliminary work for the optimization of the more complex multi–class networks presented in subsection 2.1.2.

2.2.1 A Framework for Adjoint Calculus on Networks

In this subsection we describe the setup for constrained optimization and optimal control where the constraints consist of ordinary or partial differential equations. It seems PDE–constrained optimization gained more attention in the scientific community in the 1960s, cf. [78] and the references therein. In the past decade research in the aforementioned area was strongly intensified [103, 104, 102, 99, 55, 59, 56, 58, 58].

The common structure of the optimization problems under investigation is as follows

$$\min \ J(y,u) \tag{2.107a}$$
$$\text{subject to}$$
$$c(y,u) = 0 \tag{2.107b}$$
$$u \in U_{ad} \tag{2.107c}$$

J is the so–called objective functional, $y \in Y$ is the state–variable, $u \in U_{ad} \subset U$ is an applicable control with a feasible control set U_{ad} and c is the so–called state–operator. Equation (2.107b) is known as *equation of state*.

In this subsection we review the preliminaries for and basics of PDE–constrained optimization. In particular, important concepts from functional analysis [1, 106] such as weak convergence, reflexive and dual spaces as well as adjoint operators are needed. Then we will review one method to solve problems of the form (2.107). Finally we will outline how this approach can be naturally extended to network problems.

The procedures and computations below can be made rigorous. This requires a more exhaustive treatment of function spaces and certain types of convergence. We refer to [102] for a mathematically exact treatment of optimal control of hyperbolic systems of conservation laws. Rigorous results for optimal control of elliptic or parabolic problems can be found in [78, 99, 55], for example. We note that these results do not involve PDE–constrained optimization problems on networks.

Background

Let Y, U and Z be Banach spaces. The most general form of a PDE–constrained optimal control problem is

$$\min_{y \in Y, u \in U} J(y, u) \tag{2.108a}$$

subject to

$$c(y, u) = 0, \qquad u \in U_{ad}, \qquad y \in Y_{ad} \tag{2.108b}$$

Herein, $J : Y \times U \mapsto \mathbb{R}$ is the objective function and $c : Y \times U \mapsto Z$ is the state operator. $U_{ad} \subset U$ is the set of admissible controls and $Y_{ad} \subset Y$ is the set of admissible states. Furthermore, we assume J and c to be continuously Fréchet differentiable.

We want to solve the control problem (2.108). Of course, we first need to define the notion *optimal solution* and then prove the existence of at least one such solution.

Definition 2.2.1. (Optimal solution) *A state–control pair $(y^*, u^*) \in Y_{ad} \times U_{ad}$ is called* optimal *for (2.108), if it satisfies*

$$
\begin{aligned}
c(y^*, u^*) &= 0 \\
J(y^*, u^*) &\leq J(y, u) \qquad \forall\, (y, u) \text{ satisfying } (2.108b)
\end{aligned}
$$

Compactness results are a major ingredient in existence proofs for optimal controls. The main problem is that in infinite–dimensional spaces closed and bounded sets are not compact. However, we recover compactness results if we employ the concept of weak convergence.

Theorem 2.2.2. (Weak sequential compactness [1, 106]) *Let X be a reflexive Banach space. Then the following holds:*

i) Any bounded sequence $(u_k) \subset X$ contains a weakly convergent subsequence, i.e., there are $(u_{k_i}) \subset (u_k)$ and $u \in X$ with $u_{k_i} \rightharpoonup u$.

49

ii) *Any bounded, closed and convex subset $U \subset X$ is weakly sequentially compact, i.e. any sequence $(u_k) \subset X$ contains a weakly convergent subsequence $(u_{k_i}) \subset (u_k)$ with $u_{k_i} \rightharpoonup u \in U$.*

For the rest of this subsection we need the following assumptions to hold:

1. $U_{ad} \subset U$ is convex, bounded, closed and nonempty.

2. The state equation $c(y, u) = 0$ has a continuous, bounded solution operator $u \in U_{ad} \mapsto y(u) \in Y$.

3. $(y, u) \in Y \times U \mapsto c(y, u) \in Z$ is continuous under weak convergence.

4. J is weakly sequentially lower semicontinuous and bounded from below.

Theorem 2.2.3. (Existence of an optimal control for (2.108)) *Problem (2.108) has at least one optimal solution $(y^*, u^*) \in Y_{ad} \times U_{ad}$. $u^* \in U_{ad}$ is the optimal control and $y^* = y(u^*)$ is the corresponding state.*

Proof. The proof of the theorem is relatively easy due to our assumptions; variants of the theorem with different assumptions can be found in [99].

Denote the feasible set by

$$W_{ad} \quad := \quad \{(y, u) \in Y \times U \mid (y, u) \in Y_{ad} \times U_{ad}, c(y, u) = 0\}$$

Since J is bounded from below and continuous by our general assumption and W_{ad} is nonempty, the infimum

$$J^* \quad := \quad \inf_{(y,u) \in W_{ad}} J(y, u)$$

exists and hence we find a minimizing sequence $(y_k, u_k) \subset W_{ad}$ (note that $y_k = y(u_k) \in Y_{ad}$) with

$$\lim_{k \to \infty} J(y_k, u_k) \quad = \quad J^*$$

The sequence $(u_k) \subset U_{ad}$ is bounded since by our assumption U_{ad} is bounded. Therefore we find a $\bar{u} \in U_{ad}$ with $u_{k_i} \rightharpoonup \bar{u}$. Since the solution operator for $c(y, u) = 0$ is continuous and bounded, the state sequence $y_i = y(u_{k_i})$ is bounded. Then we find a weakly convergent subsequence $(y_{i_l}) \subset Y_{ad}$ and a $\bar{y} \in Y_{ad}$ with $y_{i_l} \rightharpoonup \bar{y}$. Altogether we have a sequence $(y_{i_l}, u_{k_{i_l}}) \rightharpoonup (\bar{y}, \bar{u})$. By our third assumption we have $c(\bar{y}, \bar{u}) = 0$. By the weakly sequentially lower semicontinuity of J we then have

$$\lim_{l \to \infty} J(y_{i_l}, u_{k_{i_l}}) \quad \geq \quad J(\bar{y}, \bar{u})$$

Since $(y_{i_l}, u_{k_{i_l}})$ is a minimizing sequence and by definition of J^* we have

$$J^* = \lim_{l \to \infty} J(y_{i_l}, u_{k_{i_l}}) \quad \geq \quad J(\bar{y}, \bar{u}) \geq J^*$$

This relation shows that $(\bar{y}, \bar{u}) \in Y_{ad} \times U_{ad}$ is an optimal solution. $\qquad \square$

Note that in the proof we had to use theorem 2.2.2 twice. Uniqueness of optimal controls can in general only be guaranteed under additional assumptions on the objective functional J.

Remark 2.2.4. *Without assumptions two and three a proof for existencs of optimal controls can not be given in this generality.*

The solution operator $u \in U_{ad} \to y(u) \in Y$ for $c(y, u) = 0$ will in general not be bounded as the state operator $c : Y \times U \to Z$ can be nonlinear and involve derivatives. Additionally, the third assumption is very strong. In general, nonlinear terms do not need to be weakly continuous [1, 106, 99]. Then one needs to show strong convergence for the nonlinear terms to ensure existence of an optimal solution. Usually one then needs compact embeddings $Y \subset\subset \tilde{Y}$ to convert weak convergence in Y into strong convergence in $\tilde{Y}$; an example can be found in [78], Theorem 7.1.

Optimality conditions

In this subsection we give a necessary condition that needs to hold for an optimal solution (y^*, u^*). However, for ease of presentation, we assume $Y_{ad} = Y$ in the following, i.e., we do not consider state constraints in the sequel. More explicitly, we want to solve the following problem

$$\min_{y \in Y, u \in U} J(y, u) \tag{2.109a}$$

$$\text{subject to}$$

$$c(y, u) = 0, \quad u \in U_{ad} \tag{2.109b}$$

As before we assume $J : Y \times U \mapsto \mathbb{R}$ and $c : Y \times U \mapsto Z$ to be continuously Fréchet differentiable. We can weaken our assumptions from the previous section to some extent. We demand that the following holds:

A1 $U_{ad} \subset U$ is convex, bounded, closed and nonempty.

A2 For all $u \in U_{ad}$ the state equation $c(y, u) = 0$ has a unique solution $y = y(u)$.

A3 $c_y(y, u) \in \mathcal{L}(Y, Z)$ has a bounded inverse for all
$(y, u) \in W_{ad} := \{(y, u) \in Y \times U \mid u \in U_{ad}, c(y, u) = 0\}$

Proposition 2.2.5. *Under the stated assumptions the control–to–state mapping*

$$U_{ad} \ni u \quad \mapsto \quad y(u) \in Y$$

exists and is continuously differentiable on an open neighborhood $\tilde{U}_{ad}$ of U_{ad}.

Proof. Apply the implicit function theorem. $\qquad\square$

We continue with a definition

Definition 2.2.6. (Reduced objective functional) *The reduced objective functional $\hat{J}$ is defined by*

$$\tilde{U}_{ad} \ni u \mapsto \hat{J}(u) := J(y(u), u) \tag{2.110}$$

By proposition 2.2.5 the reduced objective funtional is well–defined and together with our assumption on J it is continuously Fréchet differentiable. Moreover, (2.109) is equivalent to

$$\min_{u \in U_{ad}} \hat{J}(u) \tag{2.111}$$

For problem (2.111) one can derive the *first order optimality condition* rather easily since it is essentially an unconstrained problem. The state equation does not appear as a constraint any more. We have the following theorem

Theorem 2.2.7. *Let $\hat{J} : U \to \mathbb{R}$ be Gâteaux differentiable and let $U_{ad} \subset U$ be nonempty and convex. If $u^* \in U_{ad}$ is an optimal solution of problem (2.111), then $u^* \in U_{ad}$ satisfies the* **variational inequality**

$$\langle \hat{J}'(u^*), u - u^* \rangle_{U^*, U} \geq 0 \quad \forall u \in U_{ad} \tag{2.112}$$

The derivative $\hat{J}'(u) \in U^*$ needs to be computed. We use a Lagrangian function which will lead to a so–called adjoint equation.

Definition 2.2.8. (Lagrangian function) *We consider (2.109). Then we define a corresponding* Lagrangian function L *by*

$$L : Y \times U \times Z^* \mapsto L(y, u, p) := J(y, u) + \langle p, c(y, u) \rangle_{Z^*, Z} \tag{2.113}$$

Consider a control $u \in U_{ad}$ and the associated state $y = y(u)$. Then the equation of state is satisfied and we have

$$\hat{J}(u) = J(y(u), u) = L(y(u), u, p) = J(y(u), u) + \langle p, c(y(u), u) \rangle_{Z^*, Z} \quad \forall p \in Z^*$$

Consequently, we formally have

$$\hat{J}'(u) = D_u L(y(u), u, p) \quad \forall p \in Z^* \tag{2.114}$$

Formally, we can compute the right hand side of (2.114). This expression can be used to derive the equation

$$\langle \hat{J}'(u), v \rangle_{U^*, U} = \langle J_y(y(u), u) + c_y(y(u), u)^* p, y'(u) v \rangle_{Y^*, Y} \tag{2.115a}$$
$$+ \langle J_u(y(u), u) + c_u(y(u), u)^* p, v \rangle_{U^*, U} \tag{2.115b}$$

where we assume here that the appearing adjoint operators are well–defined. The key idea is to choose $p \in Z^*$ such that the first term on the right hand side of (2.115) disappears. This choice gives rise to the so–called *adjoint equation*. If we have enough regularity we can state the above expression in the strong form

$$c_y(y(u), u)^* p = -J_y(y(u), u) \tag{2.116a}$$
$$\hat{J}'(u) = J_u(y(u), u) + c_u(y(u), u)^* p \tag{2.116b}$$

Equation (2.116a) is called the **adjoint equation** and (2.116b) is known as **adjoint representation** of the gradient $\hat{J}'(u)$. We summarize our results in

52

Corollary 2.2.9. (General first order necessary optimality conditions) *Let $(\bar{y}, \bar{u})$ be an optimal solution to (2.109) and let assumptions A1 to A3 hold. Then there exists an adjoint state $\bar{p} \in Z^*$ such that the following optimality conditions hold*

$$c(\bar{y}, \bar{u}) = 0 \qquad \bar{u} \in U_{ad} \qquad (2.117a)$$

$$c_y(\bar{y}, \bar{u})^* \bar{p} = -J_y(\bar{y}, \bar{u}) \qquad (2.117b)$$

$$\langle J_u(\bar{y}, \bar{u}) + c_u(\bar{y}, \bar{u})^* \bar{p}, u - \bar{u} \rangle_{U^*, U} \geq 0 \qquad \forall u \in U_{ad} \qquad (2.117c)$$

Remark 2.2.10. *Note that we need the solution of the equation of state $\bar{y} = \bar{y}(\bar{u})$ in order to compute the adjoint equation. Furthermore, we need to supply suitable initial conditions for the equations. A more rigorous and careful analysis shows that we need to solve the adjoint equation backwards in time.*

Remark 2.2.11. *The variational inequality (2.117c) can in certain situations be transformed. For example, we can take $U = L^2(\Omega)$ and $U_{ad} := \{u \in L^2(\Omega) \mid a \leq u \leq b\}$ for $a \leq b$ and identify $U^* = U$ via the Riesz–representation theorem [1, 106]. Then we have the equivalent conditions*

i) $\bar{u} \in U_{ad}$, $(\hat{J}'(\bar{u}), u - \bar{u}) \geq 0$, $\forall u \in U_{ad}$

ii) $\bar{u} \in U_{ad}$ *and*

$$\hat{J}'(\bar{u})(x) \quad \begin{cases} = 0 & if \quad a(x) \; < \; \bar{u}(x) \; < \; b(x) \\ \geq 0 & if \quad a(x) \; = \; \bar{u}(x) \\ \leq 0 & if \qquad\qquad \bar{u}(x) \; = \; b(x) \end{cases}$$

iii) $\bar{u} \in U_{ad}$ *and there are $\bar{z}_a, \bar{z}_b \in U^* = L^2(\Omega)$ with*

$$\hat{J}'(u) + \bar{z}_b - \bar{z}_a = 0$$
$$\bar{z}_a \geq 0 \qquad \bar{z}_a(\bar{u} - a) = 0$$
$$\bar{z}_b \geq 0 \qquad \bar{z}_b(b - \bar{u}) = 0$$

iv) *For any $\sigma > 0$ we have $\bar{u} = P_{U_{ad}}(\bar{u} - \sigma \hat{J}'(\bar{u}))$ with $P_{U_{ad}} = \min(\max(a, u), b)$.*

Computation of optimal controls

In general, efficient numerical methods need to rely on reformulations of (2.117c) in the spirit of remark 2.2.11. It seems computationally impossible to verify the variational inequality for all $u \in U_{ad}$ for a given $\bar{u}$. The numerical method presented in this subsection relies on a reformulation of the variational inequality as an equality in the interior of U_{ad}, cf. remark 2.2.11, ii). Generalized Newton methods are another solution technique and are based on reformulations of the variational inequality (2.117c) in the spirit of iv) in remark 2.2.11, see [100, 101]

There are two different strategies to compute an optimal control. In the first approach one derives the continuous optimality system (2.117). Then one discretizes the state and adjoint equation (2.117a) and (2.117b) as well as the relation (2.116b) for the gradient of the reduced functional. Finally one solves (2.111) by a descent type method, see [97, 69, 90]. This approach can therefore be seen as *optimize–then–discretize*. Alternatively, we can proceed by first discretizing the problem (2.109). Then we derive the optimality system for the resultant finite–dimensional optimization problem. This latter procedure is known as *discretize–then–optimize* approach.

It is well known that the negative gradient is a descent–direction; in fact, it is the steepest descent direction one can choose. Since we can evaluate the gradient, cf. (2.116b), we use a descent–method to compute the optimal solution. However, we have the constraint $u \in U_{ad}$ and in the course of the algorithm this condition might be violated. We deal with this violation by a simple projection, since our admissible set U_{ad} is typically a box. In order to guarantee convergence of the method, one has to employ a step–size rule. A suitable one is the Armijo–Goldstein rule [97, 13, 69]; it is also easy to implement. The general algorithm reads

Algorithm

1. Choose an initial control $u_0 \in U_{ad}$.

2. Compute the solution of the equation of state and the adjoint equation (2.117a) and (2.117b).

3. Compute the gradient expression (2.116b) and verify the optimality condition.

    If $\|\hat{J}'(u_0)\| = 0$ and $u_0 \in U_{ad} \setminus \bar{U}_{ad}$ then STOP.

    If $u_0 \in \bar{U}_{ad} \setminus U_{ad}$ then check the sign of the gradient, cf. ii) in remark 2.2.11.

4. Choose $\sigma > 0$ from a step–size rule and update the control.

$$\bar{u} := P(u_0 - \sigma \hat{J}'(u_0)) \in U_{ad}$$

5. Set $u_0 := \bar{u}$ and go to 2.

Extension to networks

We consider once more a graph $G = (V, A)$. On the arcs of the network we have partial differential equations and on each arc we can use the techniques from the previous subsections to derive the optimality system. However, in networks we need coupling conditions at nodes and that complicates the derivation of the adjoint system. We will see that we obtain *adjoint* coupling conditions from the original ones.

The results in this subsection are not mathematically rigorous in the sense that the appearing operators have not been studied in detail and the spaces that appear are not clearly specified. Consequently, we can not study or obtain strict existence or convergence results. From a practical point of view it is clear that there should be at least one optimal control when routing cars through a network. The numerical results indicate that the framework and the used algorithms work reliably. The formal calculations are not restricted to traffic flow networks, although we will use traffic flow as an example frequently. In the following we *assume* that the appearing operators and integral expressions are well–defined.

Let us consider one arc in a network (modeled as interval $[a, b]$) and a state $y \in Y$. We assume that the dynamics on this arc are governed by a (system of) hyperbolic conservation law(s)

$$\partial_t y + \partial_x F(y) \;=\; 0 \tag{2.118}$$

In particular, we assume F to be continuously differentiable. For a smooth test function $\phi \in C^\infty$ the corresponding weak formulation of (2.118) is then given by

$$0 \;=\; \int_0^T \int_a^b -y \partial_t \phi \, dx \, dt + \int_a^b y \phi \, \big|_0^T \, dx \, dt \tag{2.119a}$$

$$-\int_0^T \int_a^b F(y) \partial_x \phi \, dx \, dt + \int_0^T F(y) \phi \, \big|_a^b \, dt \tag{2.119b}$$

Remark 2.2.12. *In the system's case, i.e., $y \in Y^n$, products need to be read as scalar product. We have for example*

$$y \cdot \partial_t \phi \;=\; \sum_{i=1}^n y_i \partial_t \phi_i$$

We define the dual pairing

$$\langle \phi, Cy \rangle \;:=\; \int_0^T \int_a^b -y \partial_t \phi \, dx \, dt + \int_a^b y \phi \, \big|_0^T \, dx \tag{2.120a}$$

$$-\int_0^T \int_a^b F(y) \partial_x \phi \, dx \, dt + \int_0^T F(y) \phi \, \big|_a^b \, dt \tag{2.120b}$$

The (in general nonlinear) state operator is given by

$$c(y, u) \;:=\; C(y(u)) = Cy \tag{2.121}$$

Remark 2.2.13. *Note that in a network the state y will depend on controls u that are applied at certain types of junctions. Therefore, y depends on u and the notation $c(y, u)$ instead of $c(y)$ is justified.*

Remark 2.2.14. *For elliptic equations on a domain Ω one can frequently choose $Y = H_0^1(\Omega)$. If one is lucky enough (it depends on the space dimension and properties of the domain), then H_0^1 is continuously embedded in some L^p space (Sobolev embeddings). Then one can show that $Cy \in (H_0^1)^*(\Omega)$ and that the operator $c(y, u)$ is continuously Fréchet differentiable. More details can be found in [78].*

In traffic networks the controls do not enter directly in the state equations. We have the opportunity to distribute at certain nodes. Therefore, the coupling conditions take a more special form. Whenever there are $k \geq 2$ outgoing arcs at a node $v \in V$ we introduce k controls. Usually these controls have to satisfy at least one additional constraint to ensure flow conservation. Then it is computationally advantageous to eliminate controls from the model.

Consider a node $v \in V$ with $i = 1, \ldots, m$ ingoing and $j = 1, \ldots, k$ outgoing nodes. We *assume* that y has enough regularity to define a trace operator in a meaningful way, i.e., the values $y_i(b_i)$ and $y_j(a_j)$ are suitably defined. Furthermore, we assume the coupling conditions can be states as

$$y_j(a_j) \;=\; G_j(y_1(b_1), \ldots, y_m(b_m), u_1, \ldots, u_k) \tag{2.122}$$

$u_j = u_j(t)$ denotes the applicable control corresponding to outgoing road j. G_j is a (possibly nonlinear but) continuously differentiable function describing the coupling in the state and control variables.

Remark 2.2.15. *More formally, we can introduce the evaluation operator T w.r.t. x as $T_{x_0} : y(x, t) \to T_{x_0}(y(x, t)) := y(x_0, t)$ which evaluates a function of the variable x at a point x_0. However, this would just complicate the notation. We note that for the Fréchet derivative we have $T'_{x_0} \tilde{y} = \tilde{y}(x_0, t) = T_{x_0} \tilde{y}$.*

On a network $G = (V, A)$ the equation of state (2.107b) can then be written as

$$c_i(y_i, u) \;=\; 0 \qquad i = 1, \ldots, |E| \tag{2.123a}$$
$$y_j(a_j) - G_j(y_1(b_1), \ldots, y_m(b_m), u_1, \ldots, u_k) \;=\; 0 \tag{2.123b}$$
$$v \in V, |\delta^-(v)| = m, \qquad j = 1, \ldots, k, k = |\delta^+(v)|$$

Additionally, we have boundary conditions on the ingoing roads to the network and an initial condition

$$y_i(a_i, t) \;=\; g_i(t) \qquad i \text{ is an ingoing road to the network} \tag{2.124a}$$
$$y_i(x, 0) \;=\; r_i(x) \tag{2.124b}$$

Remark 2.2.16. *With the general coupling conditions (2.123b) we can easily incorporate control constraints. We demonstrate the procedure for the easiest example which occurs in traffic flow. There we have constraints of the form $\sum_{j=1}^{k} u_j = \lambda$, $\lambda \in \mathbb{R}^+$. We simply define $u_k := \lambda - \sum_{j=1}^{k-1} u_j$ to include the control constraint. We have effectively eliminated the control variable u_k which is advantageous both for the theory as well as the implementation of optimization algorithms since we satisfy the control constraint automatically!*

We *assume* that the preliminaries from the previous subsection are met, i.e., c is continuously Fréchet differentiable and the mapping $u \to y(u)$ exists. We note that we do not consider the coupling conditions (2.123b) as part of the equation of state, i.e., we do not introduce a multiplier for them in the sequel.

Theorem 2.2.17. *If all the appearing operators are well–defined, the optimality system for a network problem is given by*

$$c(y, u) = (c_1(y_1, u), \ldots, c_{|A|}(y_{|A|}, u))^T = 0^T \tag{2.125a}$$

$$c_i(y_i, u) = \partial_t y_i + \partial_x F(y_i) \tag{2.125b}$$

$$y_i(a_i, t) = g_i(t) \quad i \text{ is an ingoing road} \tag{2.125c}$$

$$y_i(x, 0) = 0 \tag{2.125d}$$

$$p(y, u) = (p_1(y_1, u), \ldots, p_{|A|}(y_{|A|}, u))^T \tag{2.125e}$$

$$\partial_t p_i + F_i'(y_i) \partial_x p_i = \partial_{y_i} J(y, u) \tag{2.125f}$$

$$p_i(x, T) = 0 \tag{2.125g}$$

$$p_j(b_j, t) = 0 \quad j \text{ is an outgoing road} \tag{2.125h}$$

The coupling conditions for the state and adjoint equations (2.125a) and (2.125e) read at a node $v \in V$ with $i = 1, \ldots, m$ ingoing and $j = 1, \ldots, k$ outgoing roads

$$y_j(a_j) = G_j(w(u), u)$$

$$F_i'(y_i(b_i))p_i(b_i) = \sum_{j=1}^{k} p_j(a_j) F_j'(G_j(w(u), u)) \partial_{x_i} G_j(w(u), u)$$

with $w := (y_1(b_1), \ldots, y_m(b_m))$. $G_j : \mathbb{R}^m \times \mathbb{R}^k \to \mathbb{R}$ is a (possibly nonlinear) differentiable function that describes how the known states and control determine the value of the state variable on the outgoing road j. The gradient w.r.t. a control u_j of the Lagrangian (2.113) is given as

$$\partial_{u_j} L(y, u, p) = \partial_{u_j} J(y(u), u) + \sum_{l=1}^{k} p_l(a) F_l'(G_l(w(u), u)) \partial_{x_{m+j}} G_l(w(u), u)$$

Remark 2.2.18. *Note that the adjoint equations need to be solved both* backwards *in time* and *on the network!*

Remark 2.2.19. *The expression $\partial_{x_i} G_j(w(u), u)$, for example, has to be understood as follows: We differentiate the function $G_j : \mathbb{R}^m \times \mathbb{R}^k \to \mathbb{R}$,*
$(x_1, \ldots, x_m, x_{m+1}, \ldots, x_{m+k}) \mapsto G_j(x_1, \ldots, x_m, x_{m+1}, \ldots, x_{m+k})$ w.r.t. x_i and then evaluate it at $(w(u), u) \in \mathbb{R}^m \times \mathbb{R}^k$.

Proof. We introduce the Lagrangian

$$L(y, u, p) = J(y, u) + \sum_{i=1}^{|A|} \langle p_i, C_i(y_i) \rangle$$

Formally, the optimality system is then still determined as

$$C_i(y_i) = 0 \tag{2.126a}$$

$$(\partial_{y_i} C_i y_i)^* p_i = -\partial_{y_i} J(y, u) \tag{2.126b}$$

$$\partial_{u_j} J(y, u) + (\partial_{u_i} C_i y_i)^* p_i = 0 \tag{2.126c}$$

57

However, the computation of $(\partial_y C_i(y_i))^*$ and $(\partial_u C_i(y_i))^*$ involves the coupling conditions (2.123b).

We use the general formula (2.114)

$$L(y(u + \tilde{u}), u + \tilde{u}, p) - L(y(u), u, p)$$

$$= J(y(u + \tilde{u}), u + \tilde{u}) - J(y(u), u) + \sum_{l=1}^{|E|} \langle p_l, (C_l(y_l(u + \tilde{u})) - C_l(y_l(u)))\rangle$$

$$= \sum_{l=1}^{|A|} \partial_{y_l} J(y, u) y'(u)\tilde{u} + \sum_{v \in V} \sum_{l=1}^{k(v)} \partial_{u_l} J(y, u)\tilde{u}_l \tag{2.127a}$$

$$+ \sum_{l=1}^{|A|} \langle p_l, (C_l(y_l(u + \tilde{u})) - C_l(y_l(u)))\rangle \tag{2.127b}$$

The summation over the controls is local to the nodes. At a node $v \in V$ we assume to have $i = 1, \ldots, m(v)$ ingoing roads and $j = 1, \ldots, k(v)$ outgoing roads. The second term needs a closer inspection. We will do so for a particular node $v \in V$. Note that only certain indices i in the above sum play a role then. We use a local labeling. As before, we consider n ingoing arcs $i = 1, \ldots, n$ and k outgoing ones $j = 1 \ldots, k$. Then we have

$$\sum_{i=1}^{n} \langle p_i, (C_i(y_i(u + \tilde{u})) - C_i(y_i(u)))\rangle + \sum_{j=1}^{k} \langle p_j(C_j(y_j(u + \tilde{u})) - C_j(y_j(u)))\rangle$$

$$= \sum_{i=1}^{n} \left(\int_0^T \int_a^b -y_i'(u)\tilde{u}\partial_t p_i \, dx \, dt + \int_a^b y_i'(u)\tilde{u}p_i \left.\right|_0^T dx \right. \tag{2.128a}$$

$$- \int_0^T \int_a^b (F_i(y_i(u + \tilde{u})) - F_i(y_i(u)))\partial_x p_i \, dx \, dt \tag{2.128b}$$

$$+ \int_0^T (F_i(y_i(u + \tilde{u})) - F_i(y_i(u)))p_i \left.\right|_a^b dt \right) \tag{2.128c}$$

$$+ \sum_{j=1}^{k} \left(\int_0^T \int_a^b -y_j'(u)\tilde{u}\partial_t p_j \, dx \, dt + \int_a^b y_j'(u)\tilde{u}p_j \left.\right|_0^T dx \right. \tag{2.128d}$$

$$- \int_0^T \int_a^b (F_j(y_j(u + \tilde{u})) - F_j(y_j(u)))\partial_x p_j \, dx \, dt \tag{2.128e}$$

$$+ \int_0^T (F_j(y_j(u + \tilde{u})) - F_j(y_j(u)))p_j \left.\right|_a^b dt \right) \tag{2.128f}$$

Since we can not control the terms $y_l'(u)\tilde{u}$, we collect corresponding terms and set them zero. This leads to the adjoint equations

$$-\partial_t p_i - F_i'(y_i)\partial_x p_i = -\partial_{y_i} J(y, u)$$
$$p_i(x, T) = 0$$

However, we need coupling conditions for the adjoint equation. We will obtain them from the expressions involving the evaluations at $x = a$ and $x = b$. They will also enable us to derive the gradient expression. On ingonig roads i we have

$$(F_i(y_i(x; u + \tilde{u})) - F_i(y_i(x; u)))p_i \big|_a^b$$

$$= F_i'(y_i(b; u))p_i(b) \sum_{l=1}^{k} \partial_{u_l} y_i(b; u) \tilde{u}_l \tag{2.129}$$

$$-F_i'(y_i(a; u))p_i(a) \sum_{l=1}^{k} \partial_{u_l} y_i(a; u) \tilde{u}_l \tag{2.130}$$

where we ignore terms involving higher than first–order derivatives in u. On outgoing roads j we need to incorporate the coupling conditions. We introduce $\tilde{w} := (y_1(b_1) + \tilde{y}_m(b_m)), \ldots, y_m(b_m) + \tilde{y}_m(b_m))$, $w := (y_1(b_1), \ldots, y_m(b_m))$ and $u := (u_1, \ldots, u_k)$. We have

$$(F_j(y_j(x; u + \tilde{u})) - F_j(y_j(x; u)))p_j \big|_a^b$$
$$= (F_j(y_j(b; u + \tilde{u})) - F_j(y_j(b; u)))p_j(b) - (F_j(y_j(a; u + \tilde{u})) - F_j(y_j(a; u)))p_j(a)$$

The first term can be linearized to yield

$$(F_j(y_j(b; u + \tilde{u})) - F_j(y_j(b; u)))p_j(b) = F_j'(y_j(b; u))p_j(b) \sum_{j=1}^{k} \partial_{u_l} y_j(b; u) \tilde{u}_l \tag{2.131}$$

The second term involves the coupling conditions which depend explicitly on the controls u. Therefore, a direct linearization as in (2.131) is not suitable. We have

$$(F_j(y_j(a; u + \tilde{u})) - F_j(y_j(a; u)))p_j(a)$$
$$= p_j(a) \left(F_j(G_j(w(u + \tilde{u}), u + \tilde{u})) - F_j(G_j(w(u), u)) \right)$$

Recall that $w(u) = (y_1(b_1; u), \ldots, y_m(b_m; u))$. Therefore we formally have two contributions to the gradient of the Lagrangian, one of which involves the partial derivatives of $y_i, i = 1, \ldots, m$. We note that $w : \mathbb{R}^k \to \mathbb{R}^n$ and $G_j : \mathbb{R}^n \times \mathbb{R}^k \to \mathbb{R}$. We formally obtain by a linearization up to first order:

$$(F_j(y_j(a; u + \tilde{u})) - F_j(y_j(a; u)))p_j(a)$$
$$= p_j(a)F_j'(G_j(w(u), u))G_j'(w(u), u) \cdot (w'(u)\tilde{u}, \tilde{u})$$

We inspect the second term a little closer. We have

$$G_j'(w(u), u) \cdot (w'(u)\tilde{u}, \tilde{u})$$

$$= \sum_{i=1}^{m} \partial_{x_i} G_j(w(u), u)(w'(u)\tilde{u})_i + \sum_{l=1}^{k} \partial_{x_{m+l}} G_j(w(u), u) \tilde{u}_l$$

$$= \sum_{i=1}^{m} \partial_{x_i} G_j(w(u), u) \left(\sum_{z=1}^{k} \partial_{u_z} y_i(b_i; u) \tilde{u}_z \right) + \sum_{l=1}^{k} \partial_{x_{m+l}} G_j(w(u), u) \tilde{u}_l \tag{2.132}$$

Clearly, the first sum will be involved to obtain the adjoint coupling conditions (due to the $y_i'(b_i; u) \tilde{u}_l$ terms) and the second one will contribute to the gradient. We want to find for a particular ingoing road i the coupling condition for the adjoint variable. By (2.129) we need to collect for $j = 1, \ldots, m$ all the terms involving $\sum_{l=1}^{k} \partial_{u_l} y_i(b; u) \tilde{u}_l$. Note that (2.132) holds on all outgoing roads $j = 1, \ldots, m$ and that we sum these expressions up according to (2.128). Then we have

$$\sum_{j=1}^{k} p_j(a) F_j'(G_j(w(u), u)) G_j'(w(u), u) \cdot (w'(u)\tilde{u}, \tilde{u})$$

$$= \sum_{j=1}^{k} p_j(a) F_j'(G_j(w(u), u)) \left[\left(\sum_{i=1}^{m} \partial_{x_i} G_j(w(u), u) \left(\sum_{z=1}^{k} \partial_{u_z} y_i(b_i; u) \tilde{u}_z \right) \right) \right.$$

$$\left. + \sum_{j=1}^{k} \sum_{l=1}^{k} \partial_{x_{m+l}} G_j(w(u), u)\tilde{u}_l \right]$$

$$= \sum_{i=1}^{m} \sum_{j=1}^{k} p_j(a) F_j'(G_j(w(u), u)) \partial_{x_i} G_j(w(u), u) \left(\sum_{z=1}^{k} \partial_{u_z} y_i(b_i; u) \tilde{u}_z \right) \qquad (2.133a)$$

$$+ \sum_{j=1}^{k} p_j(a) F_j'(G_j(w(u), u)) \sum_{l=1}^{k} \partial_{x_{m+l}} G_j(w(u), u)\tilde{u}_l \qquad (2.133b)$$

Using (2.133a) and (2.129) we obtain the following coupling condition for the adjoint equation on a particular ingoing road i

$$F_i'(y_i(b; u)) p_i(b) = \sum_{j=1}^{k} p_j(a) F_j'(G_j(w(u), u)) \partial_{x_i} G_j(w(u), u)$$

This implies that if we know $p_j(a_j)$ on the outgoing roads $j \in \delta^+(v)$, then we can compute the state on the ingoing road $i \in \delta^-(v)$. In particular, the adjoint equation needs to be solved backwards on the network. In a network the $p_j(b)$ terms are either an ingoing road at another network node $v \in V$ or belong to a terminal road. In the latter case we set $p_j(b_j) = 0$ to ensure that we do not have a contribution to the gradient of the Lagrangian function. Furthermore, in a network (2.130) will either be an outgoing node at a different junction $v \in V$ or belong to an ingoing rode at which no coupling conditions are needed.

Now we can turn our attention to collect the terms for the gradient. Again, we consider a node $v \in V$ with $j = 1, \ldots, k$ outgoing roads. Recall that we have additional contributions from the cost–functional from (2.127). We have on a road j

$$\sum_{j=1}^{k} \partial_{u_j} J(y(u), u)\tilde{u}_j + \sum_{j=1}^{k} p_j(a) \left(F_j'(G_j(w(u), u)) \sum_{l=1}^{k} \partial_{x_{m+l}} G_j(w(u), u)\tilde{u}_l \right)$$

$$= \sum_{j=1}^{k} \left(\partial_{u_j} J(y(u), u) + \sum_{l=1}^{k} p_l(a) F_l'(G_l(w(u), u)) \partial_{x_{m+j}} G_l(w(u), u) \right) \tilde{u}_j$$

Therefore we have

$$\partial_{u_j} L(y, u, p) = \partial_{u_j} J(y(u), u) + \sum_{l=1}^{k} p_l(a) F_l'(G_l(w(u), u)) \partial_{x_{m+j}} G_l(w(u), u)$$

Finally we have established all our assertions.

□

Remark 2.2.20. *For a cost functional of the form*

$$J(y,u) \;=\; \frac{1}{2}\sum_{j=1}^{|E|}\int_0^T\int_{a_j}^{b_j}\mathcal{F}_j((y_j-\bar{y}_j)^2)\,dx\,dt + \frac{\gamma}{2}\sum_{v\in V}\sum_{j=1}^{k(v)}\int_0^T u_j^2(t)\,dt$$

the appearing derivatives read

$$\partial_{y_k}J(y,u)\tilde{y} \;=\; \int_0^T\int_{a_j}^{b_j}\mathcal{F}_j'((y_j-\bar{y}_j)^2)(y_k-\bar{y}_k)\tilde{y}_k\,dx\,dt$$

$$\partial_{u_\kappa}J(y,u)\tilde{u} \;=\; \gamma\int_0^T u_{j(\kappa)}(t)\tilde{u}_{j(\kappa)}\,dt$$

We note that since $J: Y \times U \to \mathbb{R}$ we have $\partial_y J(y,u) \in \mathcal{L}(Y,\mathbb{R})$, i.e., $\partial_{y_k}J(y,u)$ is an operator that takes a function and maps it into a real number.

2.2.2 Instantaneous Control for Traffic Flow Networks

In this subsection we present an instantaneous control formulation for the optimization of traffic flow networks. There are several formulations in terms of PDEs leading to hyperbolic systems. The solution methods known so far are computationally costly; in particular for large networks, the long run-time prevents the methods from being relevant for practical applications. As the models get more complex, e.g., by considering multi–class systems, this problem becomes even more severe.

Due to the network structure we have strong coupling of the state and adjoint equations which leads to large storage requirements. Several techniques are known to adress this difficulty, among the many, we mention reduced models and snapshot techniques. In the sequel we present another approach based on instantaneous control.

The instantaneous control approach or more generally, a receding horizon strategy (RHS) [87], aims at reducing the computational effort by considering smaller time–intervals and optimize *locally* over time, see below. In general these methods will only lead to suboptimal controls for the full problem. We want to investigate how much the solutions differ in the context of our traffic flow model.

Instantaneous control problems for uncoupled systems are discussed for example in [79, 18]. Results on instantaneous control of parabolic equations and Navier-Stokes equations are reported in [55, 59, 56, 58, 57], whereas control of networks governed by the wave equation are discussed in [64]. Numerical approaches can be found in [45, 11] and the references therein.

Other approaches for efficient approximation to the solution of the optimal control problem include simplified models, see for example [33, 75]. Related is the work on optimal control for coupled systems of conservation laws to model shallow water flows as in [43, 44].

This subsection is organized as follows. Firstly, we state an optimization problem for traffic networks and present the associated adjoint system used in the optimization process [48, 51]. Secondly, we give an instantaneous control formulation and compute the corresponding adjoint system. Thirdly, we show that certain properties of the PDE model carry over to the new formulation. Furthermore, a relation between the different adjoint systems is given. A comparison of the computed controls and the run-times of the methods can be found in subsection 2.3.2.

Optimal Control Problem and Adjoint Equations

As discussed in subsection 2.1.1 we are concerned with a network $G = (V, A)$ in which the dynamics on each arc $j \in A$ are determined by the LWR–equation (cf. (2.4))

$$\partial_t \rho_j + \partial_x f_j(\rho_j) = 0$$

In the following we assume a fixed positive direction of flow, i.e., we impose the following restriction for the densities ρ_j:

$$\rho_j(x, t) < \sigma_j, \quad \forall j \in A,\ x \in [a_j, b_j],\ t \in [0, T] \tag{2.134}$$

This condition is satisfied, if we consider inflow problems to networks with suitably large maximum flux values on each road. The condition is violated in the case of backwards going shock waves.

We consider the network formulation with coupling conditions as presented in subsection 2.1.1. Recall that the network is given as a graph $G = (V, A)$. For simplicity we restrict our discussion to the case in which the network solely has the two types of junction displayed in figure 2.2: we either have one ingoing and two outgoing roads at a node $v \in V$ or two ingoing and one outgoing road. We assume that we have $K \subset V$ controls, that is K junctions of the first (or dispersing) type, cf. figure 2.2; consider the time horizon $[0, T]$ and let the objective functional be of the type

$$F(\alpha_1, \ldots, \alpha_K) = \sum_{j \in A} \int_0^T \int_{a_j}^{b_j} \mathcal{F}_j(\rho_j(x, t))\, dx\, dt + \frac{\gamma}{2} \sum_{k \in K} \int_0^T (\alpha_k(t) - \bar{\alpha}_k)^2\, dt \tag{2.135}$$

with given, smooth functions $\mathcal{F}_j : \mathbb{R} \to \mathbb{R}$ and given values $\bar{\alpha}_k$. $\gamma \geq 0$ is a weighting parameter. For notational convenience we define $\vec{\alpha}(t) := (\alpha_k(t))_{k \in K}$. We consider the optimal control problem (2.136).

$$\min_{\vec{\alpha}} F(\vec{\alpha}) \tag{2.136a}$$

$$\text{subject to } 0 \leq \alpha_k(t) \leq 1\ \forall t \in (0, T) \tag{2.136b}$$

$$\text{and such that } \rho_j(x, t) \text{ is a solution of } (2.4,\ 2.19,\ 2.20,\ 2.21)\ . \tag{2.136c}$$

Formally, the gradient $\nabla F(\vec{\alpha})[\vec{v}]$ can be determined by the adjoint system. We apply Theorem 2.2.17 with $F_j(x) = f_j(x)$ on every road j. The objective function which affects the right hand side of the adjoint equation is given by (2.135). We immediately obtain

$$\partial_t p_j(x, t) + f_j'(\rho_j(x, t)) \partial_x p_j(x, t) = \mathcal{F}_j'(\rho_j(x, t)), \quad \forall x \in [a_j, b_j],\ t \in [0, T]$$
$$p_j(x, T) = 0, \quad \forall x \in [a_j, b_j].$$

62

The boundary conditions at $x = b_j$ for p_j are given by the adjoint boundary conditions (2.137-2.139) which can be derived using Theorem 2.2.17, see remark 2.2.21 below:

B1 Boundary values for a junction k of the First Type

$$f_1'(\rho_1(b_1,t))p_1(b_1,t) = \partial_{\rho_1} u_2(\rho_1(b_1,t), \alpha_k(t))f_2'(\rho_2(a_2,t))p_2(a_2,t)$$
$$+\partial_{\rho_1} u_3(\rho_1(b_1,t), \alpha_k(t))f_3'(\rho_3(a_3,t))p_3(a_3,t)$$

or equivalently

$$p_1(b_1,t) = \alpha_k(t)p_2(a_2,t) + (1 - \alpha_k(t))p_3(a_3,t). \tag{2.137}$$

B2 Boundary values for a junction of the Second Type

$$f_1'(\rho_1(b_1,t))p_1(b_1,t) = \partial_{\rho_1} u_3(\rho_1(b_1,t), \rho_2(b_2,t))f_3'(\rho_3(a_3,t))p_3(a_3,t)$$

and

$$f_2'(\rho_2(b_2,t))p_2(b_2,t) = \partial_{\rho_2} u_3(\rho_1(b_1,t), \rho_2(b_2,t))f_3'(\rho_3(a_3,t))p_3(a_3,t)$$

or equivalently

$$p_1(b_1,t) = p_3(a_3,t), \tag{2.138a}$$
$$p_2(b_2,t) = p_3(a_3,t). \tag{2.138b}$$

B3 Boundary values for the outflow arc j, that is

$$p_j(b_j,t) = 0. \tag{2.139}$$

Since $f_j'(\rho_j) \neq 0$ by assumption (2.134), $p_j(b_j,t)$ is well-defined. The gradient $\nabla F(\vec{\alpha})[\vec{v}]$ in direction $\vec{v}$ is given by

$$\partial_{\alpha_m} F(\vec{\alpha})[v_m] = \int_0^T [\mu_s(a_s,t) - \mu_r(a_r,t)]f_m(\rho_m(b_m,t))v_m(t)dt + \tag{2.140a}$$

$$\gamma \int_0^T (\alpha_m(t) - \bar{\alpha}_m) v_m(t)dt. \tag{2.140b}$$

The adjoint equations (2.137 -2.139) are both backwards in time and in the network. This implies that the state equation (2.4) has to be solved for the complete network and all times, before it is possible to evaluate the gradient (2.140). This major complication will be adressed in the sequel by introducing instantaneous control problems. Other treatments like snapshot techniques or specialized approaches for network problems are also possible.

Remark 2.2.21. *Here we show how easy it is to obtain the coupling conditions with the aid of Theorem 2.2.17. We consider the case* B1 *first. Here we have one ingoing road at a node $v \in V$, i.e., $i = 1$ and we assume two outgoing roads, i.e., $k = 2$ in Theorem 2.2.17.*

In the following we label the ingoing road with 1 and the outgoing roads with 2 and 3, respectively. Then the coupling conditions for the equation of state read

$$\rho_2(a_2) = f_2^{-1}(\alpha_k(t)f_1(\rho_1(b))) =: G_2(\rho_1(b), \alpha_k(t)) \tag{2.141a}$$

$$\rho_3(a_3) = f_3^{-1}((1 - \alpha_k(t))f_1(\rho_1(b))) =: G_3(\rho_1(b), \alpha_k(t)) \tag{2.141b}$$

Note that the quantities in (2.141) are well-defined due to (2.134). We need to compute $\partial_{x_1}G_2$. This involves the derivative of the inverse function. We have

$$\partial_{x_1}G_2(x_1, \alpha_k(t)) = \partial_{x_1}(f_2^{-1}(\alpha_k(t)f_1(x_1)))$$

$$= \frac{1}{f_2'(f_2^{-1}(\alpha_k(t)f_1(x_1)))}\alpha_k(t)f_1'(x_1)$$

Similarly we obtain

$$\partial_{x_1}G_3(x_1, \alpha_k(t)) = \frac{1}{f_3'(f_3^{-1}((1 - \alpha_k(t))f_1(x_1)))}(1 - \alpha_k(t))f_1'(x_1)$$

By our general equation for the adjoint coupling conditions we are lead to

$$f_1'(\rho_1(b))p_1(a_1) = p_2(a)\alpha_k(t)f_1'(\rho_1(b)) + p_3(a)(1 - \alpha_k(t))f_1'(\rho_1(b)) \tag{2.142}$$

Since $f_1'(\rho_1(b)) \neq 0$ by assumption (2.134) and (2.6) we can safely cancel it in (2.142) to get the expression in B1.

For the situation in B2 we consider $i = 1, 2$ ingoing roads and label the one outgoing road with the number 3. The coupling condition for the state reads

$$\rho_3(a_3) = f_3^{-1}(f_1(\rho_1(b)) + f_2(\rho_2(b))) =: G_3(\rho_1(b), \rho_2(b))$$

We need to compute $\partial_{x_1}G_3(x_1, x_2)$ and $\partial_{x_2}G_3(x_1, x_2)$. We are led to

$$\partial_{x_1}G_3(x_1, x_2) = \left(f_3'(f_3^{-1}(f_1(x_1) + f_2(x_2)))\right)^{-1}f_1'(x_1)$$

$$\partial_{x_2}G_3(x_1, x_2) = \left(f_3'(f_3^{-1}(f_1(x_1) + f_2(x_2)))\right)^{-1}f_2'(x_2)$$

Using these results in our general formula and noting once more that we can cancle $f_i'(x_i)$ due to (2.134) and (2.6) we recover the formulas from B2.

Instantaneous Control Problem and Adjoint Equations

As already remarked, the disadvantages of the considerations presented so far are the storage requirements and the computational time. Both are due to the necessicity to solve (2.4, 2.19-2.21) and (2.140) on the whole time–interval $[0, T]$.

As usual, we derive the instantaneous control problem by discretization of the interval $[0, T]$ into N time intervals $I_i := [t_i, t_{i+1}]$, $i = 0, \ldots, N - 1$. We discretize and split the optimal control problem (2.136) according to I_i, hereby generating a sequence of optimal control problems of reduced dimension. For each interval I_i we obtain a problem determining $\vec{\alpha}(t_i)$. By piecewise constant interpolation we obtain a suboptimal solution $\vec{\alpha}(t)$ to (2.136).

We derive the instantaneous control problems as outlined above. We split the time–interval $[0, T]$ into N intervals $I_i := [t_i, t_{i+1}], i = 0, \ldots, N-1$ with $t_0 = 0, t_N = T$. We use the following notation: α_k^{i+1} is the control on the interval I_i, i.e, $\alpha_k(t) = \alpha_k^{i+1}$ for all $t \in I_i$ whereby $\alpha_k^0 := 0$. As usual, on each time interval we have k controls $\vec{\alpha}^i = (\alpha_k^i)_{k \in K}$. Additionally we define $\rho_j^i = \rho_j(t_i)$ and $\tau_{i+1} = t_{i+1} - t_i$. On the time–interval I_i we introduce the objective functional (2.143).

$$H_{i+1}(\vec{\alpha}^{i+1}) := \frac{\gamma}{4} \left(\sum_{k=1}^{K} \tau_{i+1}(\alpha_k^{i+1} - \bar{\alpha}_k)^2 + \tau_{i+1}(\alpha_k^i - \bar{\alpha}_k)^2 \right) + \qquad (2.143a)$$

$$\sum_{j=1}^{|E|} \int_{a_j}^{b_j} \frac{\tau_{i+1}}{2} \left(\mathcal{F}_j(\rho_j^{i+1}(x)) + \mathcal{F}_j(\rho_j^i(x)) \right) \, dx \qquad (2.143b)$$

Obviously, H_i is obtained by application of the trapezoid rule w.r.t. time to the functional (2.135)

$$F(\vec{\alpha}) = \sum_{i=0}^{N-1} \int_{t_i}^{t_{i+1}} \sum_{k \in K} \frac{\gamma}{2} (\alpha_k(t) - \bar{\alpha}_k)^2 + \sum_{j \in J} \int_{a_j}^{b_j} \mathcal{F}_j(\rho_j(x, t)) \, dx dt$$

The corresponding $i = 0, \ldots, N-1$ instantaneous optimal control problems on the intervals I_i are (2.144).

$$\min_{\vec{\alpha}^{i+1}} H_{i+1}(\vec{\alpha}^{i+1}) \text{ subject to} \qquad (2.144a)$$

$$0 \leq \alpha_k^{i+1} \leq 1 \qquad (2.144b)$$

$$\frac{\rho_j^{i+1}(x) - \rho_j^i(x)}{\tau_{i+1}} + \frac{\partial}{\partial x} f_j(\rho_j^{i+1}(x)) = 0 \qquad (2.144c)$$

$$\rho_j^0(x) = \rho_{j,0}(x) \qquad (2.144d)$$

and the boundary conditions as follows

A1 Boundary values for a junction k of the First Type

$$\rho_2^{i+1}(a_2) = u_2(\rho_1^{i+1}(b_1), \alpha_k^{i+1})$$
$$\rho_3^{i+1}(a_3) = u_3(\rho_1^{i+1}(b_1), \alpha_k^{i+1})$$

A2 Boundary values for a junction of the Second Type

$$\rho_3^{i+1}(a_3) = u_3(\rho_1^{i+1}(b_1), \rho_2^{i+1}(b_2))$$

A3 Boundary values for the Road s with a free boundary node at $x = a_s$:

$$\rho_s^{i+1}(a_s) = u_s^{i+1}$$

Remark 2.2.22. *Equation (2.144c) is an implicit Euler discretization in time of the conservation law. Succesful solution of (2.144) for fixed i yields controls α_k^{i+1} for $k \in K$. By piecewise constant extension we obtain $\alpha_k(t) := \alpha_k^{i+1}$ for $t \in I_i$. Hence, the instantaneous control problems are decoupled in time. However, it is not possible to solve the instantaneous control problems for $i = 0, \ldots, N-1$ in parallel. The solution on the interval I_{i+1} depends on the solution on the interval I_i.*

Equations (2.146) are the adjoint equations for the problem $i+1$ given by (2.144). They hold for all $k \in K, j \in J$ and $x \in [a_j, b_j]$ and p_j^i is the adjoint corresponding to the control $\vec{\alpha}^{i+1}$ and the state ρ_j^{i+1}.

$$\frac{p_j^i(x)}{\tau_{i+1}} - f_j'(\rho_j^{i+1}(x)) \cdot \frac{\partial}{\partial x} p_j^i(x) = -\frac{\tau_{i+1}}{2} \cdot \mathcal{F}_j'(\rho_j^{i+1}(x)) \qquad (2.146a)$$

$$p_j^i(b_j) = (p_j^i)^* \qquad (2.146b)$$

where the value of $(p_j^i)^*$ is determined by the type of junction:

B1 Boundary values for a junction k of the First Type:

$$(p_1^i)^* \cdot f_1'(\rho_1^{i+1}(b_1)) = p_2^i(a_2) \cdot f_2'(\rho_2^{i+1}(a_2)) \cdot \partial_x u_2^i(\rho_1^{i+1}(b_1), \alpha_k^{i+1})$$
$$+ p_3(a_3) \cdot f_3'(\rho_3^{i+1}(a_3)) \cdot \partial_x u_3(\rho_1^{i+1}(b_1), \alpha_k^{i+1})$$

or equivalently

$$(p_1^i)^* = \alpha_k^{i+1} \cdot p_2^i(a_2) + (1 - \alpha_k^{i+1}) \cdot p_3^i(a_3). \qquad (2.147)$$

B2 Boundary values for junctions of the Second Type:

$$(p_1^i)^* \cdot f_1'(\rho_1^{i+1}(b_1)) = p_3^i(a_3) f_3'(\rho_3^{i+1}(a_3)) \cdot \partial_x u_3(\rho_1^{i+1}(b_1), \rho_2^{i+1}(b_2))$$
$$(p_2^i)^* \cdot f_2'(\rho_2^{i+1}(b_2)) = p_3^i(a_3) f_3'(\rho_3^{i+1}(a_3)) \cdot \partial_y u_3(\rho_1^{i+1}(b_1), \rho_2^{i+1}(b_2))$$

or equivalently

$$(p_1^i)^* = p_3^i(a_3), \quad (p_2^i)^* = p_3^i(a_3). \qquad (2.148)$$

B3 Boundary values for the outflow arc j, that is

$$(p^i)_j^* = 0. \qquad (2.149)$$

Again the adjoint system is both backwards in time and backwards on the network. Of course, we can transform (2.146) into a forward system by the mapping $x \to b_j - x$. Finally, the partial derivative of H_{i+1} w.r.t. α_k^{i+1} is given by the expression

$$\partial_{\alpha_k^{i+1}} H_{i+1}(\vec{\alpha}^{i+1}) = \frac{\gamma \tau_{i+1}}{2} \left(\alpha_k^{i+1} - \bar{\alpha}_k \right) + f_1(\rho_1^{i+1}) \left(p_2^i(a_2) - p_3^i(a_3) \right) \qquad (2.150)$$

Properties of the Instantaneous Control Problem

In this part we proof properties of the instantaneous problem. To be more precise, we prove that the function $\vec{\alpha} \to H_{i+1}(\vec{\alpha})$ given by (2.143) is componentwise convex.

Theorem 2.2.23. *Let $\mathcal{F}_j$ be convex functions, f_j be concave functions. Fix a control $\vec{\alpha}^{i+1}$ and assume that the adjoint $\partial_x p_j^i$ given by (2.146a, 2.147-2.149) is non-negative for all $x \in [a_j, b_j]$.*

Then the function H_{i+1} given by (2.143) is componentwise convex as a function of $\vec{\alpha}^{i+1}$, i.e., for given $\vec{\alpha}^{i+1}$ and $\vec{\beta} := (\alpha_1^{i+1}, \ldots, \beta_k, \ldots, \alpha_K^{i+1})$ and $\beta_k \in [0, 1]$

$$H_{i+1}(\vec{\beta}) \geq H_{i+1}(\vec{\alpha}^{i+1}) + (\beta_k - \alpha_k^{i+1}) \partial_{\alpha_k^{i+1}} H_{i+1}(\vec{\alpha}^{i+1})$$

Proof. Note that $H := H_{i+1}$ does only depend on $\vec{\alpha} := \vec{\alpha}^{i+1}$ and therefore we skip the indices $i + 1$ for a moment. Further, we denote by $\rho^\beta := (\rho^\beta_j)_j$ the solution ρ^{i+1}_j to (2.144c, 2.144d, 2.145-2.146) with control $\vec{\beta}$; analogously we define ρ^α. Let $p_j := p^i_j$ be the adjoint to ρ^α and $\vec{\alpha}$.

$$
\begin{aligned}
H(\vec{\beta}) - H(\vec{\alpha}) &= \frac{\gamma\tau}{4}\left((\beta_k - \bar{\alpha}_k)^2 - (\alpha_k - \bar{\alpha}_k)^2\right) + \sum_{j=1}^{|A|} \frac{\tau}{2} \int_{a_j}^{b_j} \left(\mathcal{F}_j(\rho^\beta_j) - \mathcal{F}_j(\rho^\alpha_j)\right) dx \\
&\geq \frac{\gamma\tau}{2}(\beta_k - \alpha_k)(\alpha_k - \bar{\alpha}_k) \\
&\quad + \sum_{j=1}^{|A|} \int_{a_j}^{b_j} \frac{\tau}{2}\left(\rho^\beta_j - \rho^\alpha_j\right)\mathcal{F}'_j(\rho^\alpha_j) + p_j\left(\frac{\rho^\beta_j - \rho^\alpha_j}{\tau} + \partial_x\left(f_j(\rho^\beta_j) - f_j(\rho^\alpha_j)\right)\right) dx \\
&= \frac{\gamma\tau}{2}(\beta_k - \alpha_k)(\alpha_k - \bar{\alpha}_k) + \sum_{j=1}^{|A|} \int_{a_j}^{b_j} \left(\rho^\beta_j - \rho^\alpha_j\right)\left(\frac{\tau}{2}\mathcal{F}'_j(\rho^\alpha_j) + \frac{p_j}{\tau}\right) dx \\
&\quad \sum_{j=1}^{|A|}\left(+\int_{a_j}^{b_j} \partial_x p_j\left(f_j(\rho^\beta_j) - f_j(\rho^\alpha_j)\right) dx + \left[p_j(f_j(\rho^\beta_j) - f_j(\rho^\alpha_j))\right]_{x=a_j}^{x=b_j}\right)
\end{aligned}
$$

Since $\partial_x p_j$ is non-negative and due to $-f_j(\rho^\beta_j) + f_j(\rho^\alpha_j) \geq -(\rho^\beta_j - \rho^\alpha_j)f'_j(\rho^\alpha_j)$ we continue

$$
H(\vec{\beta}) - H(\vec{\alpha}) \geq \frac{\gamma\tau}{2}(\beta_k - \alpha_k)(\alpha_k - \bar{\alpha}) + \sum_{j=1}^{|A|}\left[p_j(f_j(\rho^\beta_j) - f_j(\rho^\alpha_j))\right]_{x=a_j}^{x=b_j}
$$

and it remains to discuss the boundary terms. Fix a junction $k_0 \in K$. Denote by m_{k_0}, r_{k_0} and so forth the connected arcs in the notation of Figure 2.2. If the junction is of the second type, then by the coupling conditions (2.146) and (2.148) and for $\vec{\gamma} \in \{\vec{\alpha}, \vec{\beta}\}$:

$$
\begin{aligned}
&-f_{r_{k_0}}(\rho^\gamma(a_{r_{k_0}}))p_{r_{k_0}}(a_{r_{k_0}}) + f_{q_{k_0}}(\rho^\gamma(b_{q_{k_0}}))p_{q_{k_0}}(b_{q_{k_0}}) + f_{p_{k_0}}(\rho^\gamma(b_{p_{k_0}}))p_{p_{k_0}}(b_{p_{k_0}}) = \\
&\left(-f_{r_{k_0}}(\rho^\gamma(a_{r_{k_0}})) + f_{q_{k_0}}(\rho^\gamma(b_{q_{k_0}})) + f_{p_{k_0}}(\rho^\gamma(b_{p_{k_0}}))\right)p_{r_{k_0}}(a_{r_{k_0}}) = 0
\end{aligned}
$$

If the junction is of the first type, we define

$$
s^\gamma := -f_{r_{k_0}}(\rho^\gamma(a_{r_{k_0}}))p_{r_{k_0}}(a_{r_{k_0}}) - f_{s_{k_0}}(\rho^\gamma(a_{s_{k_0}}))p_{s_{k_0}}(a_{s_{k_0}}) + f_{m_{k_0}}(\rho^\gamma(b_{m_{k_0}}))p_{m_{k_0}}(b_{m_{k_0}}).
$$

By the boundary conditions (2.145) and (2.147) we have

$$
\begin{aligned}
f_{r_{k_0}}(\rho^\gamma(a_{r_{k_0}})) &= \gamma_{k_0} f_{m_{k_0}}(\rho^\gamma(b_{m_{k_0}})) \\
f_{s_{k_0}}(\rho^\gamma(a_{s_{k_0}})) &= (1 - \gamma_{k_0})f_{m_{k_0}}(\rho^\gamma(b_{m_{k_0}}))
\end{aligned}
$$

and independent of $\vec{\gamma}$

$$
\alpha_{k_0} p_{r_{k_0}}(a_{r_{k_0}}) + (1 - \alpha_{k_0})p_{s_{k_0}}(a_{s_{k_0}}) = p_{m_{k_0}}(b_{m_{k_0}}).
$$

We distinguish the cases $k_0 \neq k$, where $s^\gamma = 0$ and $k_0 = k$. In the latter we note that $\rho^\alpha_{m_k} = \rho^\beta_{m_k}$ and we conclude

$$
s^\beta - s^\alpha = f_{m_k}(\rho^\alpha(b_{m_k}))(\alpha_k - \beta_k)(p_{r_k}(a_{r_k}) - p_{s_k}(a_{s_k}))
$$

Summarizing,

$$
\begin{aligned}
H(\vec{\beta}) - H(\vec{\alpha}) &\geq (\beta_k - \alpha_k)\left(\frac{\gamma\tau}{2}(\alpha_k - \bar{\alpha}) + f_{m_k}(\rho^\alpha(b_{m_k}))\,(p_{s_k}(a_{s_k}) - p_{r_k}(a_{r_k}))\right) \\
&= (\beta_k - \alpha_k)\partial_{\alpha_k} H(\vec{\alpha})
\end{aligned}
$$

$\square$

Remark 2.2.24. *The result is an extension to Theorem 2.2 in [49], where the optimal control problem (2.136) for constant controls $\vec{\alpha} \in \mathbb{R}^K$ is considered. Theorem 2.2.23 shows that the introduced instantaneous control problem (2.144) exhibits similar properties as the full problem (2.136).*

We discretize the optimality system to (2.136) in time using the same grid as before, i.e., $I_i := [t_i, t_{i+1}], i = 0, \ldots, N - 1$. For notational convenience assume that $\tau = t_{i+1} - t_i$ for all i. We apply the implicit Euler scheme for state and adjoint equations and a trapezoid rule for the gradient equation similarly to the discussion in the previous section. Let p_j^i be the adjoint to ρ_j^{i+1} for $i = 0, \ldots, N - 1$. We restrict our controls $\vec{\alpha}$ to the space of piecewise constant functions on I_i. Eventually, we obtain for $k = 1, \ldots, K$

$$
\frac{\rho_j^{i+1} - \rho_j^i}{\tau} + \partial_x f_j(\rho_j^{i+1}) = 0
$$

$$
-\frac{p_j^{i+1} - p_j^i}{\tau} + f_j'(\rho_j^{i+1})\partial_x p_j^i = \mathcal{F}_j'(\rho_j^{i+1})
$$

$$
\frac{\tau}{2}\Big((p_2^i(a_2) - p_3^i(a_3))f_1(\rho_1^{i+1}(b_1))
$$

$$
+(p_2^{i-1}(a_2) - p_3^{i-1}(a_3))f_1(\rho_1^i(b_1)) + \gamma(\alpha_k^{i+1} - \bar{\alpha}_k) + \gamma(\alpha_k^i - \bar{\alpha}_k)\Big) = 0
$$

where we skipped the boundary values, since they are exactly as in the case of the instantenous control problem, see (2.145-2.146, 2.147-2.149). To obtain a similar formulation as in the previous section we introduce the rescaled adjoint variables $p_j^i := 2/\tau p_j^i$:

$$
\frac{\rho_j^{i+1} - \rho_j^i}{\tau} + \partial_x f_j(\rho_j^{i+1}) = 0 \tag{2.151a}
$$

$$
\frac{p_j^i - p_j^{i+1}}{\tau} - f_j'(\rho_j^{i+1})\partial_x p_j^i + \frac{\tau}{2}\mathcal{F}_j'(\rho_j^{i+1}) = 0 \tag{2.151b}
$$

$$
\partial_{\alpha_k^{i+1}} H_{i+1}(\vec{\alpha}_k^{i+1}) + \partial_{\alpha_k^i} H_i(\vec{\alpha}_k^i) = 0 \tag{2.151c}
$$

Remark 2.2.25. *We observe that (2.151) has similar structure compared to the instantaneous problems given by (2.144),(2.146) and (2.150) . In (2.151) we have an additional coupling in the adjoint and gradient equation between the different time-steps i. For non-linear iterative solution schemes of (2.151) we can exploit the similarity by choosing as initial value the solution to (2.144).*

2.3 Numerical Results

In this final section related to traffic flow networks we present numerical results regarding the topics covered in the previous sections. In subsection 2.3.1 we report results on the simulation of the multi–class models discussed in subsections 2.1.2 and 2.1.4. In particular, we compare these results to results from the reference model from subsection 2.1.1. Subection 2.3.2 is devoted to the comparison of optimal controls corresponding to the reference model and the controls obtained by the instantaneous control strategy from section 2.2.2.

2.3.1 Simulations of Multi–class Models

In this subsection we give numerical results for the multi–class models discussed in sections 2.1.2 and 2.1.4. We show for some example networks that the multi–class model (2.43) indeed works as desired. In particular, we emphasize that a correct forward solver is necessary for a subsequent optimization which is not part of this thesis. The material in subsection 2.3.1 has not yet been published. More simulations are carried out for the two–class model from section 2.1.4. We report results for the situation at a simple one–one junction and on a circular network in subsection 2.3.1.

Numerical results for the multi–class model (2.43)

In this part of the work we present simulation results of the general multi–class model (2.43) for various networks. In particular we will show that the considerations regarding the choice of the controls $\alpha^i_{\varepsilon\nu}(v;t)$ are vital for a correct simulation, cf. the discussion in subsection 2.1.2.

Consider the network depicted in figure 2.17. The corresponding graph $G = (V, A)$ is given as

$$V = \{1, 2, 3, 4, 5, 6\} \qquad A = \{(1, 3), (2, 3), (3, 4), (4, 5), (4, 6)\}$$

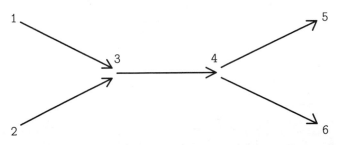

Figure 2.17: A testnetwork with one junction of the First and one junction of the Second type.

Commodity number	Starting node	Terminal node	Inflow (constant w.r.t. time)
1	1	5	γ_0^1
2	1	6	γ_0^2
3	2	5	γ_0^3
4	2	6	γ_0^4

Table 2.1: Commodities for the network from figure 2.17

Among the five roads we have two ingoing and two outgoing ones. Assume $\mathcal{S} = \{1, 2\}$ and $\mathcal{D} = \{5, 6\}$. Then we have $I = 4$ commodities, cf. table 2.1.

For ease of discussion we will consider only cases in which the inflow for both the car–density and the commodities on the ingoing roads are constant. In the sequel we will consider two different examples in which we choose different commodity inflow values $\gamma_0^i, i = 1, \dots, 4$.

We find that at node 4 all the controls are determined for the network in figure 2.17. We identify the arc $(3, 4)$ with road number 3 and the arc $(4, 5)$ with road number 4. In abuse of notation, we employ the road numbers as indices for the control variables $\alpha_{\varepsilon\nu}^i$. The analysis described in subsection 2.1.2 shows that we need to fix the controls at node 4 as follows (with the commodities defined as in table 2.1)

$$\alpha_{34}^1 = 1, \quad \alpha_{34}^2 = 0, \quad \alpha_{34}^3 = 1, \quad \alpha_{34}^4 = 0$$

In particular the control $\alpha(4; t) =: \alpha_{34}$ (which directs drivers from road 3 to road 4) is then determined by (2.35)

$$\alpha_{34} = \sum_{i=1}^{I} \alpha_{\varepsilon\nu}^i(v; t)\gamma_\varepsilon^i(b_\varepsilon) = \gamma_3^1(b_3) + \gamma_3^3(b_3)$$

Remark 2.3.1. *Note that in the reference network model (2.22) we are free to choose $\alpha_{34} \in [0, 1]$. This clearly shows that the multicommodity formulation acts as a constraint on the feasible set of controls for an optimization procedure.*

The roads are assumed to have unit length and the maximal density ρ_{max} is set to one, also. On every arc we use the flux function $f_j(\rho_j) = 4\rho_j(1 - \rho_j)$. We choose the density–inflow on the ingoing arcs $(1, 3)$ and $(2, 3)$ as

$$\rho_1(a_1, t) = 0.1 \qquad \rho_2(a_2, t) = 0.07$$

Then the influx is given as

$$f_1(\rho_1) = 0.36 \qquad f_2(\rho_2) = 0.2604$$

We run the simulation from $t = 0$ until $T = 4$ using 40 time–intervals. In particular, for the constant inflow we obtain a steady state. The solution to the LWR–equation is computed using a Godunov scheme. Once the density ρ_j is known we determine $v_j = 4(1 - \rho_j)$ and can then solve the commodity equations on road j. Then we compute new initial values for the densities and commodities using formulas (2.19), (2.20) and (2.36), respectively.

In our first test we choose

$$\gamma_0^1 = 0.3, \quad \gamma_0^2 = 0.7, \quad \gamma_0^3 = 0.8, \quad \gamma_0^4 = 0.2$$

The densities on the roads at $T = 4$ are given by

$$\rho_1 = 0.1, \quad \rho_2 = 0.07, \quad \rho_3 = 0.191942, \quad \rho_4 = 0.086575, \quad \rho_5 = 0.082891$$

and the nonzero commodity values are

$$
\begin{array}{llll}
\gamma_1^1 = 0.3, & \gamma_1^2 = 0.7, & \gamma_2^3 = 0.8, & \gamma_2^4 = 0.2, \\
\gamma_3^1 = 0.17408, & \gamma_3^2 = 0.40619, & \gamma_3^3 = 0.33578, & \gamma_3^4 = 0.08395, \\
\gamma_4^1 = 0.34143, & \gamma_4^3 = 0.65857, & \gamma_5^2 = 0.82873, & \gamma_5^4 = 0.17127
\end{array}
$$

We observe that the numerical simulation leads to $\sum_{i=1}^4 \gamma_j^i = 1$ for $j = 1, \dots, 5$. Furthermore, we have in this case

$$\alpha_{34} \;=\; \gamma_3^1 + \gamma_3^3 = 0.50986$$

We obtain for the flux on roads 3, 4 and 5

$$f_3(\rho_3) = 0.620401074544, \quad f_4(\rho_4) = 0.3163190775, \quad f_5(\rho_5) = 0.304080328476$$

We see that the flux is in principle conserved through the junctions; the minor inconsistencies are due to our rounding to the fifth digit for the density and commodity values. More importantly, we have

$$\gamma_1^1 f_1(\rho_1) = \gamma_4^1 f_4(\rho_4) \qquad \gamma_2^3 f_2(\rho_2) = \gamma_4^3 f_4(\rho_4) \qquad (2.152\text{a})$$

$$\gamma_1^2 f_1(\rho_1) = \gamma_5^2 f_5(\rho_5) \qquad \gamma_2^4 f_2(\rho_2) = \gamma_5^4 f_5(\rho_5) \qquad (2.152\text{b})$$

This means precisely that the flow of cars corresponds to the stated commodities and the model works as desired. Note that by the assumption $\gamma_1^1 = 0.3$ we require 30 % of the drivers starting at node 1 to arrive at node 5. The quantity $\gamma_4^1 f_4(\rho_4)$ states that $\gamma_4^1 \cdot 100$ % of the flux on road 4 corresponds to commodity 1, i.e., started at node 1 and arrived at node 5. The equality $\gamma_1^1 f_1(\rho_1) = \gamma_4^1 f_4(\rho_4)$ states that all the drivers that want to go from node 1 to node 5 indeed arrive at node 4. Therefore, the model works as desired.

In the second example we leave the densities on the ingoing roads unchanged, yet we modify the initial commodity values which of course changes the downstream density and commodity values. We compute for the densities

$$\rho_3 = 0.191942, \quad \rho_4 = 0.098802, \quad \rho_5 = 0.071118$$

Note that the value on road 3 is the same as before; this is not surprising as we have a merging junction and by conservation of the flux we should also get the same value for the density. Additionally we obtain for the nonzero commodity values

$$
\begin{array}{llll}
\gamma_1^1 = 0.7, & \gamma_1^2 = 0.3, & \gamma_2^3 = 0.4, & \gamma_2^4 = 0.6, \\
\gamma_3^1 = 0.40619, & \gamma_3^2 = 0.17408, & \gamma_3^3 = 0.16789, & \gamma_3^4 = 0.25184, \\
\gamma_4^1 = 0.70755, & \gamma_4^3 = 0.29245, & \gamma_5^2 = 0.40872, & \gamma_5^4 = 0.59128
\end{array}
$$

Then we have for the control from the reference network (2.22)

$$\alpha_{34} = \gamma_3^1 + \gamma_3^3 = 0.57408$$

which differs from the value for the first example! Again equations (2.152) hold with the new values for the fluxes on roads 4 and 5.

These two examples provide evidence that the model and its implementation are correct.

We conclude this section with a slightly more involved example to show that we can indeed run simulations on larger networks.

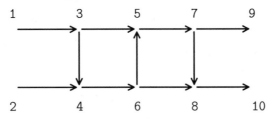

Figure 2.18: A testnetwork with three junctions of First and Second type.

We consider the network displayed in figure 2.18. We assume that the sources and destinations are given as $\mathcal{S} = \{1, 2\}$ and $\mathcal{D} = \{9, 10\}$ and we have again 4 commodities, cf. table 2.2.

Commodity number	Starting node	Terminal node	Inflow (constant w.r.t. time)
1	1	9	0.7
2	1	10	0.3
3	2	9	0.6
4	2	10	0.4

Table 2.2: Commodities for the network from figure 2.18

We use the preprocessing routine described in appendix B to determine which of the variables $\alpha_{\varepsilon\nu}^i$ need to be fixed a priori. We find that for the network in figure 2.18 out of the 12 control variables $\alpha_{\varepsilon\nu}^i$ we have to fix 8 a priori at the three dispersing junctions for the four commodities occuring in this network. Therefore, we solely have 4 controls that can be chosen:

$$\alpha_{(1,3),(3,4)}^1, \qquad \alpha_{(1,3),(3,4)}^2, \quad \alpha_{(4,6),(6,5)}^2, \quad \alpha_{(4,6),(6,5)}^4$$

One can check that these are the controls that can be variable. For example, we require $\alpha_{(4,6),(6,5)}^1 = 1$ since otherwise some drivers belonging to commodity 1 with destination 9 will not be able to get there.

Remark 2.3.2. *In the reference network model (2.22) we would have 3 variables, one at each dispersing junction.*

We choose a constant inflow

$$\rho_1(a_1, t) = 0.2 \qquad \rho_2(a_2, t) = 0.1$$

on the ingoing roads 1 and 2 (corresponding to arcs $(1,3)$ and $(3,4)$, respecitvely). Together with the constant inflow for the commodity values, cf. table 2.2, we expect to converge to a steady state. Again, we choose the flux function $f_j(\rho_j) = 4\rho_j(1-\rho_j)$. Our simulations are conducted on the time–interval $[0,4]$ choosing $\Delta t = 0.1$. Δx is chosen such that the CFL–condition

$$1 > \frac{\Delta t}{\Delta x} \max_{\xi \in [0,1]} f'_j(\xi) = \frac{\Delta t}{\Delta x} 4$$

is satisfied. In figures 2.19 – 2.21 we compare the results of two different simulations. In the first one we choose the variable controls as follows:

$$\alpha^1_{(1,3),(3,4)} = 0.2, \qquad \alpha^2_{(1,3),(3,4)} = 0.3, \quad \alpha^2_{(4,6),(6,5)} = 0.75, \quad \alpha^4_{(4,6),(6,5)} = 0.5$$

In the second simulation we define them as

$$\alpha^1_{(1,3),(3,4)} = 0.7, \qquad \alpha^2_{(1,3),(3,4)} = 0.8, \quad \alpha^2_{(4,6),(6,5)} = 0.1, \quad \alpha^4_{(4,6),(6,5)} = 0.2$$

We observe that this change affects the density and commodity distribution in the steady state we converge to. As an example, we give results for the arc $(6,5)$ for the network from figure 2.18 in figures 2.19 – 2.21.

Numerical results for the two class model

In this part we focus on the computation of solutions to Riemann–problems for the two–class model from subsection 2.1.4 and use them to compute a solution for a one–one junction and a ring–road. All the following results have been obtained by a second–order relaxed scheme, see for example [67]. At the intersection the states are determined as described in subsection 2.1.4. The results presented are obtained with a discretization width of $\Delta x = 1/800$ on the roads of a network $G = (V, A)$.

First, we consider a situation of two connected roads with different free flow velocities c^j, cf. equation (2.89). The intersection is located at $x^1_v = x^2_v = 1/2$. We consider constant initial data on both roads, i.e.,

$$U^1_0 = \begin{pmatrix} 0.2 \\ 0.1 \end{pmatrix}, \qquad U^2_0 = \begin{pmatrix} 0.4 \\ 0.1 \end{pmatrix}, \tag{2.153}$$

and free flow velocities

$$c^1 = 0.8 \text{ and } c^2 = 1. \tag{2.154}$$

In figure 2.22 we present a contour plot of the flux $c^j \rho^j v^j(\rho^j), j = 1, 2$. As required by our coupling conditions, the flux is continuous through the road intersection. In figure 2.23 we present the time–evolution of the densities ρ^j_1 and ρ^j_2 on both roads $j = 1, 2$. Due to the low inflow $\rho^1 v^1(\rho^1)$, the constant state U^1_0 is preserved on the

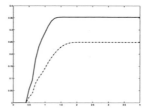

Figure 2.19: Density on arc $(6,5)$ for the network from figure 2.18. The solid line corresponds to the first, the dash–dotted line to the second choice for the variable controls.

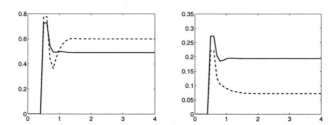

Figure 2.20: Values for commodities 1 (left) and 2 (right) on arc $(6,5)$ for the network from figure 2.18. The solid line corresponds to the first, the dash–dotted line to the second choice for the variable controls.

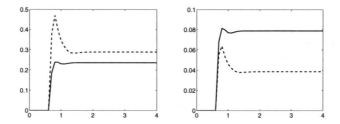

Figure 2.21: Values for commodities 3 (left) and 4 (right) on arc $(6,5)$ for the network from figure 2.18. The solid line corresponds to the first, the dash–dotted line to the second choice for the variable controls.

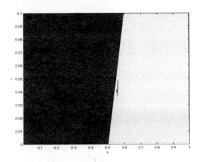

Figure 2.22: Flux for two connected roads with intersection at $x = 1/2$.

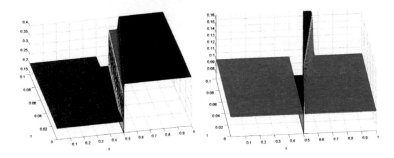

Figure 2.23: Densities ρ_1 and ρ_2 for two connected roads with different free flow velocities

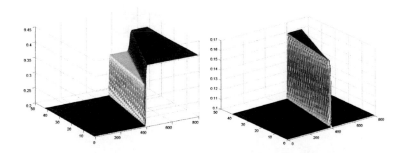

Figure 2.24: Densities ρ_1 and ρ_2 for two connected roads with equal free flow velocities.

incoming road $j = 1$. On the outgoing road $j = 2$, we observe a composition of a 1- and a 2-wave. Neither ρ_1 nor ρ_2 has to be continuous through the intersection. For comparison we present the solution $U^j(x,t), j = 1, 2$ in the case

$$c^1 = c^2 = 1, \tag{2.155}$$

in figure 2.24. Similarly, we observe a composite wave on the outgoing road $x = 1/2$. The presented solution is in this case equivalent to the solution to a (standard) Riemann problem with initial data given above. Finally, note that our scheme resolves the arising contact discontinuity well.

Next, we present results on the network represented in figure 2.25, i.e., with two roads $j = 1, 2$. The circular network has an inflow arc $j = 2$ at $x = x_0$ where we produce a time-varying inflow in both the ρ_1 and ρ_2 density. At first this generates waves on the part of the road $x > x_0$. These waves now travel along the road and reach after some time ($\Delta t \approx 0.0152$) again the intersection at $x = x_0$. Now, the combined inflow of the circular road ($j = 1$) and the additional road, exceeds the maximal possible flow. Hence, we observe a shock wave forming and moving on $x < x_0$. This pattern can be seen in all subsequent pictures. Moreover, we apply a slightly different condition at the intersection $x = x_0$: Instead of requiring an equal flux condition (as in theorem 2.1.31) we propose the following: If the sum of the demands $d_1(\rho^1(x_0+,t)) + d_2(\rho^2(x_0,t))$ does not exceed the supply $s_1(\rho^1(x_0+,t))$, then we allow all the flow to pass the intersection. Otherwise, we impose (2.104). This condition is as in [51] and also allows for a unique solution of (2.105), see [51, 19].

We depict the numerical results for the circular road $j = 1$ only. In this case, the drawing area is such that $x = 0$ and $x = 1$ correspond to the same point in the network graph from figure 2.25. The intersection is located at $x = x_0 = 1/2$ and we set $c^1 = c^2 = 1$. The first picture (figure 2.26) shows a contour plot on road 1 of the flux $Q(x,t) := \rho^1(x,t)v(\rho^1(x,t))$ for $0 \leq x \leq 1$. The discontinuity in $Q(x,t)$ at $x = x_0$ is due to the fact that $Q(x_0+,t) = Q(x_0-,t) + (\rho^2(x_0,t)v(\rho^2(x_0,t)))$. Along $(x_0,t), t < 0.0152$, we see the influence of the inflow on road $j = 2$ leading to increasing and decreasing total flux. Furthermore, we observe a backwards moving shock wave after $\Delta t = 0.0152$. Additionally, we plot the density $\rho^1(x,t)$ in figure 2.26 where the shock wave can be seen also. For completeness, we also give the results on contour plots of the densities ρ_1^1 and ρ_2^1 on road $j = 1$ in figure 2.27.

2.3.2 Evaluation of the Instantaneous Control Strategy

In this subsection, we first review an optimization algorithm which we use in our numerical computations. Then we compare the results of the full optimal control problem (2.136) to the ones of the instantaneous control approach, i.e., the consecutive solution of (2.144) on each time-interval I_i. In our tests we use different networks and various parameter settings. We compare both the computational time and the final optimal controls obtained with the different control strategies. Finally, we present results on a real-world example.

In order to solve the sequence of optimization problems (2.144), we use a gradient-projection method as presented in [17, 88, 89] combined with a scaled version of the steepest descent algorithm, see for example [69, 97]:

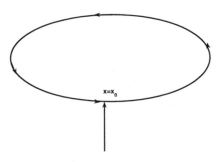

Figure 2.25: Sketch of the circular network.

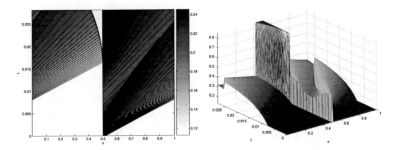

Figure 2.26: Flux $\rho(x,t)v(\rho(x,t))$ and $\rho(x,t)$ on the circular road $i = 1$.

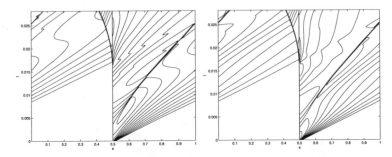

Figure 2.27: Densities $\rho_1^1(x,t)$ (left) and $\rho_2^1(x,t)$ (right) on the circular road $j = 1$.

Algorithm. (2.156)

1. For each instantaneous control problem $i = 0, \ldots, N-1$

2. Solve the state equation (2.144c) with control α^{i+1} and obtain for each $j \in A$ by piecewise constant extension $\rho_j^{i+1}(x)$ on the rectangle $[a_j, b_j] \times [t_i, t_{i+1}]$.

3. Solve the adjoint equation (2.146) on time–interval I_i and obtain μ_j^i for each $j \in J$ and the gradient $\nabla H_{i+1}(\vec{\alpha}^{i+1})$ by the gradient equation (2.150).

4. Compute a step of the gradient projection method with a stepsize β obtained by the Goldstein-Armijo-rule [97] and a diagonal matrix D with positive entries to be specified below.

$$\vec{\alpha}^{i+1} := P(\vec{\alpha}^{i+1} - \beta D \nabla H_{i+1}(\vec{\alpha}^{i+1})).$$

Herein, P is the projection on the feasible set $[0,1]^K$.

5. Repeat steps 2-4 on the time–interval I_i until convergence and then go to the next problem $i+1$

In our test we choose the matrix D, such that $\|D\nabla H_{i+1}(\vec{\alpha}^{i+1})\|_\infty = const$, other choices such as BFGS-type methods are possible [69]. Of course, step 4 and the iteration in step 5 of the above algorithm are costly – even in the instantaneous case. Therefore, Hinze et. al. introduced in [56, 57, 55] a "one–gradient step" technique for practical applications. This can be seen as skipping step 5 in algorithm (2.156) and thereby computing step 4 only once for each subproblem. We will refer to this algorithm as "one–step" in the numerical results presented below. Furthermore, the stepsize β is chosen as a constant in [57]. Optimal controls of high quality at very cheap computational costs are reported in [57, 18] and references therein. We stress the fact, that in our applications the main point is the storage requirement, since the underlying dynamics is governed by a scalar conservation law only.

To compare the results from the instantaneous control approach with those from problem (2.136), we solve (2.136) as follows:

We choose piecewise constant control functions $k \in K$:

$$\alpha_k = \sum_{i=0}^{N-1} c_k^{i+1} \chi_{I_i}$$

where χ_{I_i} is the characteristic function on I_i and $c_k^{i+1} \in [0,1] \subset \mathbb{R}$. We discretize (2.4) by an implicit Euler scheme in time and (2.135) by the trapezoid rule and obtain the set of equations (2.151). Furthermore, we use standard ODE–solvers, i.e., explicit Runge–Kutta–methods [12], to compute the functions $\rho_j^i(x), \mu_j^i(x)$. The optimization–algorithm is similar to the one presented above. For a fixed control $t \to \vec{\alpha}(t)$ we solve state (2.4), adjoint (2.137) and gradient (2.140) equation. The new iterate is obtained by gradient projection with the Goldstein-Armijo stepsize rule, similar to (2.157). We iterate until the convergence criterion is reached.

We consider the following test–cases for comparison between the optimal and the instantaneous control problems. As in [85, 19] we use the concave flux function

$$f_j(\rho_j) = f(\rho_j) = 4\rho_j(1 - \rho_j)$$

on all arcs $j \in A$. The function $\mathcal{F}_j$ appearing in the objective functional (2.135) is given by [75]

$$\mathcal{F}_j(\rho_j) = \rho_j$$

and if not stated otherwise we set $\gamma = 0$ in (2.135). Initially, all networks are empty, i.e., $\rho_j(x, 0) = 0$. Additionally, we introduce $L_j := b_j - a_j$ as length of the roads in the network. The concrete lengths for the testcases are specified later. The optimization method is terminated in both cases, if either the normalized difference in the functional values is less than 10^{-4} or if the l_1–norm of the projected gradient is less than 10^{-4}. The geometry of the networks is as in figures 2.28 - 2.29. The network on the left of figure 2.28 is used to compare the qualitative behavior of the optimal controls to both the full optimal control problem (2.136) and the instantaneous problem (2.144). For computations on this network we set $L_j = 1$. The network to the right models the major

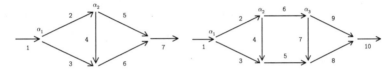

Figure 2.28: Two road networks: to the left a test network, to the right a "real-world" example.

highways from Frankfurt to Munich (Germany). The lengths correspond to the kilometers on the various highways, i.e.,

$$(L_1, \dots, L_{10}) = (1, 217, 119, 148, 93, 11, 175, 39, 270, 1).$$

A third testcase is given by the set of networks depicted in figure 2.29 and used to compare the computational costs. $K/2$ nodes in the top and bottom row are controllable by independent controls.

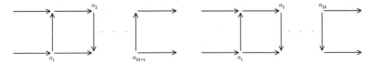

Figure 2.29: A scalable network

We present results for different traffic situations. We extend the solution to the instantaneous control problem by a constant to a function on $[0, T]$ and denote the optimal control by $\alpha_v^*(t)$ for the junction $v \in V$. By N we denote the number of intervals I_i, i.e. the number of instantaneous control problems. First, we consider the network on the left of figure 2.28 and prescribe a constant inflow $\rho_1(a_1,t) = 1/4$. We expect the optimal control in both cases to be $\alpha^*(t) = (\alpha_1^*(t), \alpha_2^*(t)) = (\frac{1}{2}, 1)$, since then the flow is distributed equally on the shortest paths from road 1 to road 7. Numerically, we observe that $\alpha_2^*(t) \equiv 1$ for both models and α_1^* is given in figure 2.30. The results are similar for different values of N and the optimal control differs on the order of 10^{-3} for both models in this case. For further comparisons we introduce the function

$$\mathcal{H}(t) := \sum_{j \in A} \int_{a_j}^{b_j} \mathcal{F}_j(\rho_j(x,t)) dx \tag{2.157}$$

For $\mathcal{F} = id$ and the test–network with constant inflow $\rho_1(a_1,t)$ we formally calculate that

$$\partial_t \mathcal{H} = f_1(\rho_1(a_1,t)) - f_7(\rho_7(b_7,t)).$$

Due to the initial data, we note that $\partial_t \mathcal{H} \geq 0$ and for T sufficiently large $\partial_t \mathcal{H} \to 0$. This implies that $\mathcal{H}$ is monotone increasing and reaches an asymptotic value.

In figure 2.30 we now can compare the effect of instantanenous, one–gradient step (i.e., skipping step 5 of algorithm (2.156)) and the optimal control (2.136). As expected, the solution to (2.136) is superior to the other approaches, but the solution to (2.144) obtained by algorithm (2.156) and by the one–gradient step modification of (2.156) are of similar type and close to the optimal solution.

Next, we consider the "real-word" network with $N = 25$ time–intervals. In this case, the inflow is chosen as a step-function in order to model the more realistic situation of peak-periods followed by periods of lower traffic volume. The optimal controls are given in figure 2.31 and we observe that the controls qualitatively differ due to the decoupling (in time) of the control problems in the instantaneous case.

Furthermore, the instantaneous control problems can only optimize with respect to the actual current inflow and do not take into account the behaviour of inflow profiles in the future. More elaborate receding horizon strategies incorporate more time–intervals to alleviate this effect; details can be found in [87].

A further comparison of computational costs is given below for various network sizes with a geometry as in figure 2.29 and different numbers of time–intervals. We choose three networks with $|A| = 11$, 17 and 32 respectively corresponding to 3, 5 and 10 control functions $\alpha_v(t)$. For each number of arcs and controls in the network we conducted computations varying the parameter N and using constant inflow profiles $\rho_1(a_1,t) = \rho_2(a_2,t) = 0.2$ on the two ingoing roads. We also investigated different time-discretization schemes, e.g. an equidistant ($\tau_i = const$) one and logarithmic ones. We found that the qualitative behavior of the controls and the computational time does not depend on the choice of τ_i.

Increasing the number of time-intervals leads to an increase in the computation times for the optimal control problem (2.136) and to an increased number of instantaneous problems (2.144). We report CPU-times for solving the full optimal control problem (2.136) and the N instantaneous optimal control problems (2.144) in table 2.3. We also report the average number of Armijo steps necessary to determine β in step 4 and the number of projected gradient steps used in step 5 of the algorithm. Note that in the case of the one–step gradient methods the CPU–time for the instantaneous problem will be

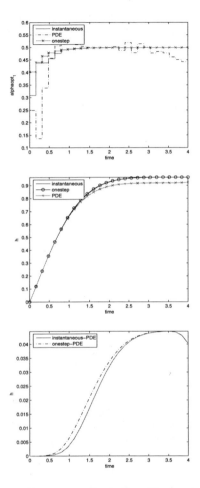

Figure 2.30: In all plots results corresponding to the optimal control to (2.136) are denoted by "PDE", to the control obtained by solving with algorithm (2.156) by "instantaneous" and the one–step modification of algorithm (2.156) by "one–step", resepectively. Optimal controls $\alpha_1^*(t)$ for $N = 25$ (top), plot of the function $\mathcal{H}$ given by equation (2.157) (middle) and difference of instantaneous and one–step afer normalization (bottom).

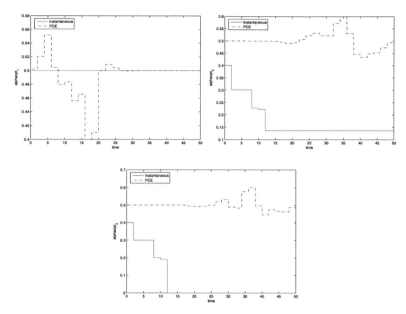

Figure 2.31: Optimal controls $\alpha_1^*(t)$ (top), $\alpha_2^*(t)$ (middle) and $\alpha_3^*(t)$ (bottom) for real-world network with $N = 25$ time-intervals for the optimal control problem (2.136) ("PDE") and N instantaneous problems (2.144) ("instantenous").

reduced again by a factor of $10 - 15$ corresponding to the average number of projected gradient steps. Using a fixed step–width (see [57]) instead of the Armijo–rule the solution of (2.144) is computed by solving the state and adjoint equation only once.

As the results show, in most cases the instantaneous approach is 3 to 4 times faster than the full model. The CPU–time improvement rises up to a factor of 30 to 60 if we compare the full PDE model with the one–step gradient method. Another advantage of the instantaneous control approach is the reduced dimension of the problems to solve: The solution to state and adjoint equation and the gradient expression are needed only for a single time–step. Hence, the required storage is reduced by a factor N.

# Arcs		N = 10	N=20	N=30	N=40	N=50
	Full OCP	66.5	146.2	344.3	378.3	565.1
	Gradient steps	20	20	26	22	22
$|A| = 11$	Av. Armijo tests	1	1.2	2.15	1.59	2.36
	Instantaneous OCP	75.7	135.1	154.9	172.4	228.9
	Av. gradient steps	21.9	16.05	9.57	7.03	6.48
	Av. Armijo tests	1.08	1.27	1.56	1.47	1.82
	Onestep OCP	3.6	6.7	11.1	17.1	20.8
	Av. Armijo tests	1	1	1	1	1.02
	Full OCP	108.6	292.6	546.8	615.3	829.1
	Gradient steps	22	27	25	23	24
$|A| = 17$	Av. Armijo tests	1	2	2.64	1.82	1.75
	Instantaneous OCP	123.0	273.2	262.812	227.2	275.3
	Av. gradient steps	24.8	25.8	14.0	7.1	5.8
	Av. Armijo tests	1.3	1.2	1.3	1.63	1.82
	Onestep OCP	4.6	9.8	15.5	23.1	32.3
	Av. Armijo tests	1	1	1	1.08	1.22
	Full OCP	356.3	427.2	1081.3	1061.3	1549.4
	Gradient steps	36	23	31	25	28
$|A| = 32$	Av. Armijo tests	1.89	1.74	2.48	1.24	1.28
	Instantaneous OCP	228.4	388.4	505.7	363.2	366.8
	Av. gradient steps	26.6	21.4	17.8	7.85	5.16
	Av. Armijo tests	1.09	1.14	1.15	1.55	2.40
	Onestep OCP	8.6	21.8	29.7	42.5	60.2
	Av. Armijo tests	1	2	1.37	1.3	1.42

Table 2.3: CPU-times in seconds. CPU–time needed to solve (2.136), number of gradient steps (step 5) and average number of Armijo–tests per gradient step needed to find an acceptable stepsize. CPU–time needed to solve **all** N instantaneous control problems (2.144), average number of gradient steps per each instantaneous control problem and average number of Armijo tests per gradient iteration and per instantaneous problem.

Chapter 3

Supply Networks

Different approaches for the simulation of continuous supply chain models using partial differential equations have been introduced during the last years; see for example [4, 3, 5, 29]. For the purpose of this presentation, we are interested in supply chain models for networks which are mainly derived in [4] and extended in [38, 39]. The latter includes the formulation of coupling conditions at intersections by introducing time-dependent queues governed by the mass-flux in the network. In particular, the flow of parts can be controlled at the vertices of the network.

In section 3.1 we present one particular continuous model for supply networks. For this model we derive the optimality system with the methods presented in subsection 2.2.1. This enables us to apply our optimization procedure presented in the context of taffic networks.

In section 3.2 we give some results on the computation of optimal controls for supply networks. Additionally, we compare the computed optimal controls to results form an associated discrete optimization problem.

3.1 Optimization of Supply Networks

An important aspect in supply chain decision making are optimization problems, for example maximizing output of a production process or minimizing used buffers. In our model the constraints involve partial and ordinary differential.equations which need to be discretized for a numerical treatment.

For our supply network model this discretization can be chosen, such that the optimization problem is in fact a mixed–integer programming problem, see [32] and subsection 3.1.1 below. This is mainly due to the fact that the governing dynamics in the supply network are linear in the state (but not in the control) variables. In [32] further extensions to the mixed–integer problem have been investigated, e.g., finite size buffers, inflow profile optimization or processor shut–down due to maintenance.

3.1.1 Adjoint Calculus for Supply Networks

In this subsection we show how the general framework from subsection 2.2.1 can be used to solve a certain class of optimization problems for supply networks.

We briefly review our supply network formulation. Then we derive the continuous optimality system and show that the mixed–integer formulation is also a valid discretization of the discretized continuous optimality system, i.e., both approaches *discretize–then–optimize* and *optimize–then–discretize* briefly described in subsection 2.2.1 lead to the same continuous optimal control if the discretization width tends to zero. We defer the numerical comparison of the presented approaches to subsection 3.2.2.

Modeling supply networks

We briefly recall the continuous supply network model. For more details we refer to [38, 39]. In the following we consider a directed graph $G = (V, A)$ consisting of a set of arcs A and a set of vertices V. Each arc corresponds to one processor (or supplier). The length L^e of one processor is mapped on the interval $[a^e, b^e]$. For a fixed vertex $v \in V$, the set of ingoing arcs is denoted by δ_v^- and the set of outgoing arcs by δ_v^+. For notational convenience we set $x_v^e := a^e$, if $e \in \delta_v^+$, and $x_v^e := b^e$, if $e \in \delta_v^-$. The maximal processing capacity μ^e and the processing velocity v^e of each processor are constant parameters on each arc. According to the assumption that each processor possesses a queue, we locate a queue at the vertex $v \in V$ in front of the processor. In the case of more than one outgoing arc, we introduce distribution rates $A^{v,e}(t)$, $v \in V_d$ where $V_d \subset V$ denotes the set of dispersing junctions. Those rates describe the distribution of incoming parts among the outgoing processors and are later subject to optimization. The functions $A^{v,e}$ are required to satisfy $0 \leq A^{v,e}(t) \leq 1$ and $\sum_{e \in \delta_v^+} A^{v,e}(t) = 1$ for all times $t > 0$.

The continuous supply network model consists of a coupled system of partial and ordinary differential equations. Here, the transport inside each processor $e \in A$ is governed by a simple advection equation:

$$\partial_t \rho^e + \partial_x f^e(\rho^e) = 0, \tag{3.1}$$

where

$$f^e(\rho^e) = v^e \rho^e \tag{3.2}$$

for some given velocity v^e and where the density of the parts is given by ρ^e on each arc $e \in A$.

As technical detail we need to introduce boundary data for those arcs $e \in A$ which are incoming to the network, i.e., such that $\delta_v^- = \emptyset$. Here, we assume inflow data $\rho_0(t)$ to be given and set $\rho^e(a, t) = \rho_0(t)$ for all $v \in V$ and $e \in \delta_v^+$ and $\delta_v^- = \emptyset$. From now on we neglect this technical point.

Whenever a processor is connected to another processor of possibly different maximal capacity μ^e, we introduce a buffering zone for the incoming but not yet processed parts. To describe the buffering we introduce the time–dependent function $q^e(t)$ describing the load of the buffer or queue. The dynamics of the buffering is governed by the difference of all incoming and outgoing parts at the connection point: If the queue is empty, the outgoing flux is either a percentage of the sum of all incoming fluxes given by $A^{v,e}(t)$ or the maximal processing capacity. In the first case the queue remains empty, in the second case the queue increases. Last, if the queue is full, the queue is always reduced with a capacity determined by the distribution rates $A^{v,e}$ and the capacities of the connected arcs. Mathematically, the situation can be described by the following ODE:

$$\partial_t q^e(t) \;=\; A^{v,e}(t) \sum_{\bar{e}\in\delta_v^-} f^{\bar{e}}(\rho^{\bar{e}}(x_v^{\bar{e}},t)) - f^e(\rho^e(x_v^e,t)) \tag{3.3}$$

$$f^e(\rho^e(x_v^e,t)) \;=\; \begin{cases} \min\{A^{v,e}(t)\Big(\sum_{\bar{e}\in\delta_v^-} f^{\bar{e}}(\rho^{\bar{e}}(x_v^{\bar{e}},t))\Big),\mu^e\} & q^e(t)=0 \\ \mu^e & q^e(t)>0 \end{cases} \tag{3.4}$$

Optimal control problems for supply networks

For further investigation we apply the following modifications and simplifications: First, in order to avoid the the discontinuous dependence on the queue-length in (3.4), we make use of the reformulation presented in [2]. There, equation (3.4) has been replaced

$$f^e(\rho^e(x_v^e,t)) = \min\{\mu^e, \frac{q^e(t)}{\varepsilon}\} \quad \text{with } \epsilon \ll 1. \tag{3.5}$$

See [2] for further remarks. Since adjoint calculus requires the constraints to be differentiable, we replace the function $y \to \min(y/\epsilon, \mu^e)$ in (3.5) by any smooth approximation $\psi^{e,\delta}(y)$ for the computations following. To be more precise, we assume there are families of smooth functions $\{\psi^{e,\delta}\}$ such that

$$\lim_{\delta\to 0} \psi^{e,\delta}(y) = \min(y/\epsilon, \mu^e) \;\forall y, \forall e. \tag{3.6}$$

For notational convenience we drop the superindex δ in the following, since the calculations remain true for all $\delta > 0$. Third, we simplify the notation by introducing functions $h^e(\vec{\rho}, \vec{A}^{v,e})$ for each edge e (resp. $\tilde{e}$) and fixed $v \in V$ such that $e \in \delta_v^+$ (resp. $\tilde{e} \in \delta_v^+$). We define

$$h^e(\vec{\rho}^e, \vec{A}^v) = A^{v,e}(t) \sum_{\bar{e}\in\delta_v^-} f^{\bar{e}}(\rho^{\bar{e}}), \forall e \in \delta_v^+ \setminus \{\tilde{e}\}, \tag{3.7a}$$

$$h^{\tilde{e}}(\vec{\rho}^e, \vec{A}^v) = \left(1 - \sum_{e\neq\tilde{e}} A^{v,e}(t)\right) \sum_{\bar{e}\in\delta_v^-} f^{\bar{e}}(\rho^{\bar{e}}). \tag{3.7b}$$

Note that with this definition in the spirit of remark 2.2.16 the assumption $\sum_{e\in\delta_v^+} A^{v,e} = 1$ can be omitted. For example, for an intersection with $\delta_v^- = \{1\}$ and $\delta_v^+ = \{2,3\}$ (corresponding to a dispersing junction or a junction of First Type in the traffic context) we have the more explicit form

$$h^2(\vec{\rho}^e, \vec{A}^v) = A^{v,2}(t)f^1(\rho^1), \; h^2(\vec{\rho}^e, \vec{A}^v) = (1 - A^{v,2}(t))f^1(\rho^1). \tag{3.8}$$

Finally, we introduce a measure for the performance of the supply chain network. A general cost functional is given for example by (3.9a). This particular choice of the cost functional aims at the minimization of the size of queues and the number of parts in the network. However, other choices are possible; in subsection 3.2.2 we present an example in which we just want to maximize the output of a particular supply network.

Summarizing, (3.9) constitutes a constrained optimal control problem where the constraints are given by linear transport and ordinary differential equations.

$$\min_{A^{v,e}(t), v \in \mathcal{V}_d} \sum_{e \in A} \int_0^T \int_{a^e}^{b^e} f^e(\rho^e(x,t)) \, dx \, dt + \int_0^T q^e(t) \, dt \tag{3.9a}$$

subject to

$$e \in A, \ v \in V, \ t \in (0,T), \ x \in [a^e, b^e] \tag{3.9b}$$

$$\partial_t \rho^e(x,t) + v^e \partial_x \rho^e(x,t) = 0 \tag{3.9c}$$

$$\partial_t q^e(t) = h^e(\vec{\rho}^e, \vec{A}^v) - \psi^e(q^e), \ q^e(0) = 0. \tag{3.9d}$$

The vector of controls $\vec{A}^v := (A^{v,e})_{e \in \delta_v^+}$ contains the individual distribution rates $A^{v,e}$ and the dependent states are collected in the vector $\vec{\rho}^e := (\rho^e)_{e \in A}$. In the sequel we are concerned with the numerical solution to the previous optimal control problem.

Derivation of Optimality Systems for the Optimal Control Problem

In the sequel we make the optimal control problem (3.9) accessible to certain solution techniques. Different approaches are possible. In [32] the optimal control problem has been solved by reformulating it as mixed–integer model. This is possible, if one introduces a coarse grid discretization of (3.9). Below we will derive a discrete optimality system for this discretization and – contrary to [32] – solve the latter directly by nonlinear optimization methods. This approach is known as *first discretize then optimize*. Formally, one can also derive the continuous optimality system and discretize the latter. This method is referred to as *first optimize then discretize*; we present the corresponding results below. Furthermore, the relation between the approaches *first discretize then optimize* and *first optimize then discretize* will be given for the optimal control problem (3.9).

Optimality System of the Discrete Optimal Control Problem

First, we consider the discrete optimality system. A coarse grid discretization in space of (3.9c) is obtained simply by a two-point upwind discretization and (3.9d) is discretized usind the explicit Euler method. Each arc has length L^e and we introduce a step size Δt such that the CFL condition for each arc and the stiffness restriction of the ordinary differential equation are met. The time steps t_j are numbered by $j = 0, \ldots, T$. We use the following abbreviations for all e, j :

$$\rho_j^{e,b} := \rho^e(b^e, t_j), \ \rho_j^{e,a} := \rho^e(a^e, t_j), \ q_j^e := q^e(t_j), A_j^{v,e} := A^{v,e}(t_j) \tag{3.10}$$

$$h_j^e := h^e(\vec{\rho}^e(x, t_j), \vec{A}^v(t_j)). \tag{3.11}$$

Due to the boundary condition $v^e \rho^e(a,t) = \psi^e(q^e(t))$ we replace the discrete variable $\rho_j^{e,a}$ by $\psi^e(q_j^e)$ and therefore, $\rho_j^{e,a}$ does not appear explicitly in the discrete optimal control problem below. For the initial data we have

$$\rho_0^{e,b} = \rho_0^{e,a} = q_0^e = 0, \ \forall e. \tag{3.12}$$

Finally, the discretization of problem (3.9) reads for $j \geq 1, e \in A, v \in V$:

$$\min_{\vec{A}^v,\, v\in V_d} \sum_{e\in A} \sum_{j=1}^{T-1} \Delta t \left(\frac{L^e}{2}(\psi^e(q_j^e) + v^e \rho_j^{e,b}) + q_j^e \right) \tag{3.13a}$$

subject to

$$\rho_{j+1}^{e,b} = \rho_j^{e,b} + \frac{\Delta t}{L^e}(\psi(q_j^e) - v^e \rho_j^{e,b}) \tag{3.13b}$$

$$q_{j+1}^e = q_j^e + \Delta t(h_j^e - \psi^e(q_j^e)) \tag{3.13c}$$

For deriving the discrete optimality system we state the precise definition of h^e in the case of the following intersections, see Figure 3.1.

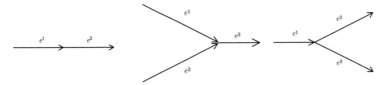

Figure 3.1: Sample intersections labeled as type I, II and III (from left to right)

I h^2 is independent of $\vec{A}^v$ and we have $h^2(\vec{\rho}^x, \vec{A}^v) = v^1 \rho^1(b,t)$.

II We obtain $h^3(\vec{\rho}^x, \vec{A}^v) = v^1 \rho^1(b,t) + v^2 \rho^2(b,t)$.

III As already stated, we have in the controlled case III:

$$h^2(\vec{\rho}^x, \vec{A}^v) = A^{v,2}(t) v^1 \rho^1(b,t) \quad \text{and} \quad h^3(\vec{\rho}^x, \vec{A}^v) = (1 - A^{v,2}(t)) v^1 \rho^1(b,t).$$

Remark 3.1.1. *The previous system (3.13) for $\delta = 0$ can be reformulated as mixed–integer problem [32] by adding binary variables to reformulate the relation $v^e \rho^e(x_v^e, t) = \min\{\mu^e, q^e(t)/\epsilon\}$: let ζ_j^e be a binary variable, i.e., $\zeta_j^e \in \{0;1\}$. Then,*

$$v^e \rho_j^{e,a} = \min\{\mu^e, q^e(t)/\epsilon\}$$

is equivalent to

$$\mu^e \zeta_j^e \ \leq \ v^e \rho_j^{e,a} \leq \mu^e$$

$$\frac{q_j^e}{\epsilon} - M\zeta_j^e \ \leq \ v^e \rho_j^{e,a} \leq \frac{q_j^e}{\epsilon}$$

$$\mu^e \zeta_j^e \leq \ \frac{q_j^e}{\epsilon} \ \leq \mu^e(1 - \zeta_j^e) + M\zeta_j^e$$

for $M > 0$ sufficiently large. Indeed, the value ζ_j^e is determined by the relation of μ^e and q_j^e/ϵ. E.g., if $\mu^e > q_j^e/\epsilon$, then we obtain $\zeta_j^e = 1$. Since both formulations are equivalent, we derive the discrete optimality system by considering system (3.13).

Now it is straightforward to derive the discrete optimality system for (3.13). In the spirit of the proof of Theorem 2.2.17 we denote the Lagrange multipliers for the discretized partial differential equation by λ_j^e and for the discretized ordinary differential equation by p_j^e. The discrete Lagrangian is given by

$$L(\vec{\rho}_j^e, \vec{q}_j^e, \vec{A}_j^v, \vec{\lambda}_j^e, \vec{p}_j^e) = \sum_{e \in \mathcal{A}} \sum_{j=1}^{T-1} \Delta t \left(\frac{L^e}{2} (\psi^e(q_j^e) + v^e \rho_j^{e,b}) + q_j^e \right) - \tag{3.14a}$$

$$\sum_{e \in \mathcal{A}} \sum_{j=1}^{T} \Delta t L^e \lambda_j^e \left(\frac{\rho_{j+1}^{e,b} - \rho_j^{e,b}}{\Delta t} - \frac{\psi(q_j^e) - v^e \rho_j^{e,b}}{L^e} \right) - \tag{3.14b}$$

$$\sum_{e \in \mathcal{A}} \sum_{j=1}^{T} \Delta t \, p_j^e \left(\frac{q_{j+1}^e - q_j^e}{\Delta t} - (h_j^e - \psi(q_j^e)) \right), \tag{3.14c}$$

if we set $\lambda_T^e = p_T^e = 0$. Assuming sufficient constraint qualifications the first–order optimality system is given by equations (3.13c) and (3.13b) and the following additional equations for $j \leq T, e \in A$ and $v \in V$:

$$\lambda_{j-1}^e = \Delta t \frac{v^e}{2} + \lambda_j^e - \frac{\Delta t}{L^e} \left(\phi_j^e - v^e \lambda_j^e \right), \tag{3.15a}$$

$$\phi_j^e := \sum_{\bar{e} \in \delta_v^+ \text{ s.t. } e \in \delta_v^-} p_j^{\bar{e}} \frac{\partial}{\partial \rho^e} h_j^{\bar{e}}, \tag{3.15b}$$

$$p_{j-1}^e = \Delta t (1 + \frac{L^e}{2} (\psi^e)'(q_j^e)) + p_j^e - \Delta t \left(p_j^e - \lambda_j^e \right) (\psi^e)'(q_j^e), \tag{3.15c}$$

$$0 = \sum_{e \in \delta_v^+} p_j^e \frac{\partial}{\partial A^{v,\bar{e}}} h_j^e \tag{3.15d}$$

The summation in the definition of the function ϕ^e is understood in the following way: For a fixed intersection $v \in V$ such that $e \in \delta_v^-$ we sum over all $\bar{e} \in \delta_v^+$. Hence, the function ϕ^e depends on the type of intersection and for clearity we state its explicit form for the cases I–III introduced above:

I: We have $\phi_j^2 = 0$ and $\phi_j^1 = p_j^1 v^1$.

II: We obtain $\phi_j^1 = p_j^3 v^3$ and $\phi_j^2 = p_j^3 v^3$.

III: For this more interesting case we find $e = 1$ which implies
$\phi_j^1 = A^{v,2} p_j^2 v^2 + (1 - A^{v,2}) p_j^3 v^3$.

Furthermore, we obtain with the previous definitions for $\bar{e} \neq \tilde{e}$:

$$\sum_{e \in \delta_v^+} p_j^e \, \partial_{A^{v,\bar{e}}} h_j^e = \left(p_j^{\bar{e}} - p_j^{\tilde{e}} \right) \sum_{e \in \delta_v^-} v^e \rho_j^e. \tag{3.16}$$

Summarizing, the optimality system to (3.13) is given by (3.13c,3.13b) and (3.15). Changing the objective functions only affects the first term on the right hand side in formulas (3.15a) and (3.15c).

Optimality system of the continuous optimal control problem

In this part we turn our attention to the continuous optimality system for (3.9); we will show that the optimality system (3.13c), (3.13b) and (3.15) from subsection 3.1.1 is a valid discretization of the former. For the derivation of the continuous optimality system to (3.9) the Lagrangian reads

$$L(\vec{\rho}^e, \vec{A}^v, \vec{q}^e, \vec{\Lambda}^e, \vec{P}^e) = \sum_{e \in A} \int_0^T \int_{a^e}^{b^e} v^e \rho^e \, dx dt + \int_0^T q^e dt - \tag{3.17a}$$

$$\sum_{e \in A} \int_0^T \int_{a^e}^{b^e} \Lambda^e \partial_t \rho^e + \Lambda^e v^e \partial_x \rho^e \, dx dt - \tag{3.17b}$$

$$\sum_{e \in A} \int_0^T P^e \left(\partial_t q^e - h^e(\vec{\rho}^e, \vec{A}^v) + \psi^e(q^e) \right) dt \tag{3.17c}$$

In this setup the adjoint variables are denoted as $\Lambda^e(x, t)$ and $P^e(x, t)$; we use captial letters to highlight their difference from the previously introduced quantities λ_j^e and p_j^e. The relation between these variables is discussed below. With a similar technique as in the proof of theorem 2.2.17 we formally obtain the continuous optimality system for all $t, x \in [a^e, b^e], e \in A$ as

$$\partial_t \rho^e + v^e \partial_x \rho^e = 0, \ \rho^e(x, 0) = 0, \ v^e \rho^e(a, t) = \psi^e(q^e), \tag{3.18a}$$

$$\partial_t q^e = h^e(\vec{\rho}^e, \vec{A}^v) - \psi^e(q^e), \ q^e(0) = 0, \tag{3.18b}$$

$$-\partial_t \Lambda^e - v^e \partial_x \Lambda^e = v^e, \ \Lambda^e(x, T) = 0, \tag{3.18c}$$

$$v^e \Lambda^e(b, t) = \sum_{\bar{e} \in \delta_v^+ \text{ s.t. } e \in \delta_v^-} P^{\bar{e}}(t) \frac{\partial}{\partial \rho^{\bar{e}}} h^{\bar{e}}(\vec{\rho}^e, \vec{A}^v), \tag{3.18d}$$

$$-\partial_t P^e = 1 - \left(P^e - \Lambda^e(a, t) \right) (\psi^e)'(q^e), \ P^e(T) = 0, \tag{3.18e}$$

$$\sum_{e \in \delta_v^+} P^e \frac{\partial}{\partial A^{v, \bar{e}}} h^e(\vec{\rho}^e, \vec{A}^v) = 0. \tag{3.18f}$$

Recall that in the limit case $\delta = 0$, we have by definition $\psi^e(y) \to \min\{y/\epsilon, \mu^e\}$. Therefore, we obtain

$$(\psi^e)'(q^e) \to \frac{1}{\epsilon} H(\mu^e - q^e/\epsilon), \ \delta \to 0, \tag{3.19}$$

where $H(x)$ is the Heaviside function. Hence, in the limit the dynamics of the adjoint queue P^e is governed by a discontinuous right–hand side.

We conclude this subsection with a result on the relation of the different optimality systems. We show that the approaches *first discretize then optimize* and *first optimize then discretize* are equivalent in the sense that we converge towards the continuous (and correct) Lagrangian multipliers. More precisely, we have the following

Theorem 3.1.2. *The multiplier of the discrete optimality system (3.15), (3.13b), (3.13c) converges to the continuous multiplier of (3.18) if the discretization widths L^e and Δt tend to zero. Therefore, the discrete optimality system is a suitable discretization of its continuous counterpart.*

Proof. We reformulate the discrete optimal control problem in the introduced variables defined by

$$\Lambda_j^{e,a} := \lambda_j^e - \frac{L^e}{2}, \quad P_j^e := p_j^e, \tag{3.20}$$

Then, (3.15),(3.13c),(3.13b) read

$$\frac{\rho_{j+1}^{e,b} - \rho_j^{e,b}}{\Delta t} = -\frac{v^e}{L^e}(\rho_j^{e,b} - \rho_j^{e,a}), \ \rho_0^e = 0, \ v^e \rho_j^{e,a} = \psi^e(q_j^e), \tag{3.21a}$$

$$\frac{q_{j+1}^e - q_j^e}{\Delta t} = h_j^e - \psi^e(q_j^e), \ q_0^e = 0 \tag{3.21b}$$

$$\frac{\Lambda_{j-1}^{e,a} - \Lambda_j^{e,a}}{\Delta t} = v^e - \frac{v^e}{L^e}\left(\Lambda_j^{e,b} - \Lambda_j^{e,a}\right), \ \Lambda_T^{e,a} = 0, \tag{3.21c}$$

$$v^e \Lambda_j^{e,b} = \sum_{\bar{e} \in \delta_v^+ \text{ s.t. } e \in \delta_v^-} p_j^{\bar{e}} \frac{\partial}{\partial \rho^e} h_j^{\bar{e}}, \tag{3.21d}$$

$$\frac{P_{j-1}^e - P_j^e}{\Delta t} = 1 - \left(P_j^e - \Lambda_j^{e,a}\right)(\psi^e)'(q_j^e), \tag{3.21e}$$

$$0 = \sum_{e \in \delta_v^+} P_j^e \frac{\partial}{\partial A^{v,\bar{e}}} h_j^e \tag{3.21f}$$

Obviously, (3.21) is an upwind and explicit Euler discretization of (3.18). Note that the discrete Lagrangian multiplier λ_j^e and the discretized Lagrange multiplier $\Lambda_j^{e,a}$ satisfy

$$\Lambda_j^{e,a} = \lambda_j^e + \mathcal{O}(L^e) \tag{3.22}$$

and L^e is in fact the discretization stepwidth in space. Therefore, if we formally let $L^e, \Delta t \to 0$ for $\frac{L^e}{\Delta t}$ fixed, we see that $\lambda^e \to \Lambda^e$ and furthermore, the discrete Lagrangian tends to the continuous Lagrangian. $\square$

3.2 Numerical Results

In this section we present numerical results concerning the quality of approximations and computation times of the approaches presented in section 3.1.

In subsection 3.2.1 we test our adjoint–based formulas for the gradient. Then we give a mixed–integer programming (MIP) formulation for the continuous supply network model. This enables us to compare the optimal controls resulting from our adjoint approach to the ones from the MIP model. Finally, we conduct a study on the computational times for the two different formulations.

3.2.1 Gradient computations

At first we compare the gradient of the cost functional obtained by finite differences to the gradient obtained by the adjoint equations for a suitable network. We use the network depicted in figure 3.2 for this test since it has only two variable controls $A_j^{1,2}$ and $A_j^{2,6}$ at time j (recall that $A_j^{1,3} = 1 - A_j^{1,2}$ and $A_j^{2,5} = 1 - A_j^{2,6}$ due to the coupling conditions). We discretize the square $[0,1] \times [0,1]$ using 16 points in each direction. Therefore we have $16 \cdot 16 = 256$ points $(A_j^{1,2}, A_j^{2,6})$ at which we compare the gradient from the adjoint equation to the expression from the finite differences (3.23), cf. figures 3.4 – 3.6.

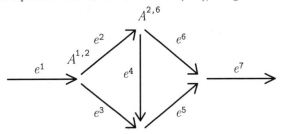

Figure 3.2: Sample network with controls $A_j^{1,2}$ and $A_j^{2,6}$

We set the time–horizon $T = 4$, use $NT = 200$ time–intervals and set $\epsilon = 1$. We use a one–sided forward difference scheme to compare the gradient at time–interval j, $j = 1, \ldots, NT$:

$$\partial_{A_j^{v,e}} J(\vec{A}^v) \quad := \quad \frac{J(\vec{A}^v + \delta) - J(\vec{A}^v)}{\delta} \tag{3.23}$$

where $\delta = 0.001$. For the cost–functional we choose the nonlinear function

$$J(\vec{A}^v) \quad := \quad \left(\sum_{e \in A} \sum_j \Delta t \left(\frac{L^e}{2} (\psi(q_j^e) + v^e \rho_j^{e,b}) + q_j^e \right) \right)^2 . \tag{3.24}$$

Furthermore, we set $L^3 = L^6 = 10$ and $L^e = 1$ for $e \in A\{e^3, e^6\}$. The processing rates are $\mu^e = 1$, $\forall e \in A$. This implies that the lowest functional value should be attained for $A_j^{1,2} = 1$ and $A_j^{2,6} = 0$ for all j as confirmed in figure 3.3. The inflow–profile on e^1 is chosen as

$$f^{in}(t) \quad = \quad \begin{cases} 0.852 & t \leq 2 \\ 0 & t > 2 \end{cases}$$

With this inflow profile the gradient w.r.t. $A_{50}^{1,2}$ is nonzero and is depicted in figure 3.4. The controls $A_j^{1,2}$ with $j > 2$ can be chosen arbitrarily since the inflow is zero and hence the gradient w.r.t. these controls needs to vanish. However, since in this particular setup queue 2 is nonempty at time $j = 200$, the gradient w.r.t. $A_{200}^{2,6}$ does not vanish, cf. figure 3.5. The relative error in the second component is of order $1e^{-8}$ and can be found in figure 3.6.

Finally, we mention that we have conducted extensive test with different objective–functionals and varying the parameters $\epsilon \in [0.01, 1]$, $\delta \in \{1e^{-2}, 1e^{-3}, 1e^{-4}, 1e^{-5}\}$ and $NT \in [20, 400]$; we never encountered a relative error in the gradient larger than $1e^{-6}$.

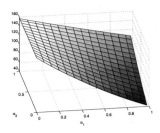

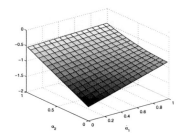

Figure 3.3: Plot of the cost functional (3.24) corresponding to Figure 3.2.

Figure 3.4: First component of the gradient computed by the adjoint scheme at $j = 50$

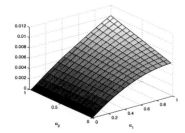

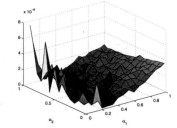

Figure 3.5: Second component of the gradient computed by the adjoint scheme at $j = 200$

Figure 3.6: Relative error in partial w.r.t. $A_{200}^{2,6}$ at $j = 200$

Remark 3.2.1. *If we choose $L^e \equiv 1$ in the test–network from figure 3.2 we make the following observation: If the inflow is low enough such that there is no backup in the queues, i.e., the processors can always process all the input the queue supplies, the controls do not play any role. Consequently, the gradient should be zero which we can indeed verify with our algorithm. Furthermore, this fact serves as a validation for the correctness of our adjoint formulas and gradient expressions.*

3.2.2 Optimal Controls for a Supply Network

In fact, there are two different approaches for solving the optimal control problem given in (3.13b), (3.13c) and (3.15). On the one hand, we use a steepest descent method for a suitable cost functional. We consecutively solve the equations of state (3.13b) and (3.13c) for a given initial control $\vec{A}_0^v \equiv 0$ and the adjoint equations (3.15a)–(3.15c) which in turn are needed to evaluate the gradient (3.16). Using the Armijo–Goldstein rule for the choice of the stepsizes we update the controls $\hat{A}_0^v$ and iterate the described procedure. This is similar to algorithm (2.156), solely the equations under consideration need to be changed.

On the other hand, we reformulate (3.13b), (3.13c) and (3.15) as a mixed-integer programming (MIP) model. The continuous optimal control probelm (3.9) is discretized using a two point upwind scheme for the PDE and a explicit Euler method for the ODE (see [32] for some details). The proposed MIP model is stated as follows:

$$\min \sum_{e \in A} \sum_{j=1}^{T-1} \Delta t \left(\frac{L^e}{2} (v^e \rho_j^{e,a} + v^e \rho_j^{e,b}) + q_j^e \right) \tag{3.25a}$$

subject to

$$\rho_{j+1}^{e,b} = \rho_j^{e,b} + \frac{\Delta t}{L^e}(v^e \rho_j^{e,a} - v^e \rho_j^{e,b}) \tag{3.25b}$$

$$\sum_{e \in \delta_v^-} v^e \rho_j^{e,b} = \sum_{e \in \delta_v^+} h_j^e \tag{3.25c}$$

$$q_{j+1}^e = q_j^e + \Delta t(h_j^e - v^e \rho_j^{e,a}) \tag{3.25d}$$

$$\mu^e \zeta_j^e \leq v^e \rho_j^{e,a} \leq \mu^e \tag{3.25e}$$

$$\frac{q_j^e}{\varepsilon} - M\zeta_j^e \leq v^e \rho_j^{e,a} \leq \frac{q_j^e}{\varepsilon} \tag{3.25f}$$

$$\rho_j^{e,a}, \rho_j^{e,b}, h_j^e, q_j^e \geq 0, \tag{3.25g}$$

$$\zeta_j^e \in \{0,1\}, \tag{3.25h}$$

with $e \in A$, $j = 1, \ldots, T$, and M a sufficiently large constant. The essential difference to (3.13b) is to rewrite the nonlinearity in (3.5) by introducing binary variables ζ_j^e. This leads finally to a mixed-integer problem and not simply to a linear programming (LP) model. We use the standard optimization software solver ILOG CPLEX [30] to compute the solution to the mixed–integer problem (3.25).

Quality of solutions of discrete adjoint calculus compared with the mixed-integer model

As a next step, we compare results computed by the adjoint approach and the mixed-integer programming (MIP) model presented in (3.25). We show that the approaches induce the same optimal functional value for a particular class of cost functionals. We choose a cost functional that aims at maximizing the output of a particular supply network.

In the following, we consider the network in Figure 3.7. It consists of 11 processors and queues and we have the six free controls $A^{2,3}(t)$, $A^{2,4}(t)$, $A^{2,5}(t)$, $A^{2,6}(t)$, $A^{2,7}(t)$ and $A^{9,10}(t)$.

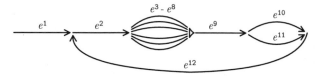

Figure 3.7: Sample network

The artificial arc 1 is used to prescribe an inflow profile which is given by

$$
f(t) \;=\;
\begin{cases}
0.5 & 0 \;\le\; t \;\le\; \frac{T}{4} \\[4pt]
0.1 & \frac{T}{4} \;<\; t \;\le\; \frac{T}{2} \\[4pt]
0.3 & \frac{T}{2} \;<\; t \;\le\; \frac{3}{4}T \\[4pt]
0 & \frac{3}{4}T \;<\; t \;\le\; T
\end{cases}
\tag{3.26}
$$

Our goal is to maximize the output of processor 12 on a given time–interval $[0, T]$. We use an equidistant time–discretization with NT time–intervals and choose the following rather simple cost functional

$$
J(\vec{A}^v) \;=\; \sum_{j=2}^{NT+1} -\frac{v^{12}\rho_j^{12,b}}{j}.
\tag{3.27}
$$

In the example below we define $T = 200$, $NT = 400$, $\epsilon = 1$ and set $L^e = v^e = 1$ for all edges except for $e = 2$; here we use $L^2 = 1$ and $v^2 = 2$. The corresponding processing rates are given in Table 3.1.

e	1	2	3	4	5	6	7	8	9	10	11	12
μ^e	100	8	10	0.5	0.5	10	0.5	2	20	3.5	2.5	8

Table 3.1: Processing rates μ^e

In figures 3.8 – 3.12 we present results for the optimal routing problem by pointing out similarities and differences between the adjoint and discrete (MIP) approach. The computation of the adjoint approach takes $37.781s$ using 32 iterations in algorithm (2.156); for the MIP we obtain a computation time of $16.60s$ using 16 iterations in ILOG CPLEX [30]. In the adjoint approach, we terminate the iteration if the relative error of two consecutive iterates is less than $tol := 1e^{-6}$ - consistent with the default accuracy in ILOG CPLEX [30]. Both approaches yield an optimal functional value of $J^*(\vec{A}^v) = -6.49$.

In figure 3.8 we plot the optimal outflow profile computed by the two approaches. We observe that for this particular example the curves coincide. However, the computed optimal controls and time evolution of the queues differ considerably. In figures 3.9 and 3.10 we plot the optimal control feeding parts into queue 10. Furthermore, we present the evolution over time for the queue 10 in figures 3.11 and 3.12 for the MIP and the adjoint approach, respectively. We observer that the maximum queue–length for queue 10 is slightly higher in the adjoint computations than in the MIP solution, cf. figures 3.11 and 3.12.

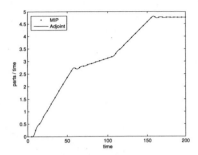

Figure 3.8: Optimal output for processor 12 over time.

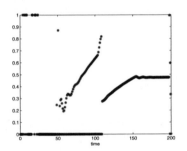

Figure 3.9: Plot of the distribution rate $A_j^{9,10}$ computed by the MIP.

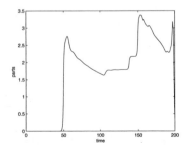

Figure 3.10: Plot of the distribution rate $A_j^{9,10}$ computed by the adjoint approach.

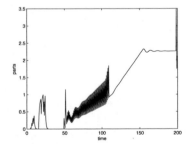

Figure 3.11: Optimal queue length q_j^{10} computed by the MIP.

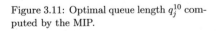

Figure 3.12: Optimal queue length q_j^{10} computed by the adjoint approach.

Since the optimal functional values coincide we see that we do not have a unique minimizer to our optimal control problem.

Computational times

The numerical results concerning supply networks conclude with a comparison of computational times of the adjoint–based approach and the mixed-integer formulation. Our computations are performed on the network given in Figure 3.7 with default parameters $v^e = L^e = 1, e = 2, \ldots, 12$, $\epsilon = 1$ and time horizon $T = 200$. To obtain a stable discretization both models have to satisfy the following restriction:

$$\Delta t \leq \min\{\epsilon; \frac{L^e}{v^e} : e \in A\}. \tag{3.28}$$

Resulting from (3.28) the parameter NT describes the number of time intervals. We increase NT by varying the ratio of L^1/v^1. The MIP is solved using the interior point method implemented in ILOG CPLEX [30].

NT	Adjoint	MIP
200	7.31	5.52
400	26.10	17.06
800	45.10	68.09
2000	124.58	592.61

Table 3.2: CPU times in sec for sample network Figure 3.7

As Table 3.2 indicates the MIP is superior if one wants to use up to approximately 600 time-steps (corresponding to $\Delta t \in [0.\overline{3}, 1]$). As NT increases the adjoint approach becomes more attractive. For values of $\Delta t < 0.\overline{3}$ it computes an optimal solution faster than the MIP. At present the MIP fails to compute a solution for $\Delta t \leq 0.05$ since the the system becomes too large and the preprocessing procedure produces infeasible solutions.

Chapter 4

Kinetic Models for Traffic Flow

This chapter deals with a different approach for traffic flow modeling. We consider a single road with multiple lanes. One can model each lane with a macroscopic model. However, incorporating driver interactions such as braking, acceleration and passing one another can hardly be modeled properly. Therefore, we consider mesoscopic or kinetic models for traffic flow [66, 52].

Kinetic models are based on a statistical description of the drivers' behavior. Consequently, we model the evolution of a density function $f(t, x)$ on a road. This leads to Boltzmann– [74] and Vlasov–Fokker–Planck–type [66, 50] equations. These models mostly differ in the way interaction of drivers is accounted for. Different braking and acceleration terms are used as well as different expressions for the passing probability.

In section 4.1 we derive the well–known Aw–Rascle model [7] from a kinetic model equation. The derivation shows that certain terms arising from the kinetic formulation need to be neglected to arrive at the Aw–Rascle model. This indicates that certain kinetic models contain more information than standard second–order models. Therefore, we have another incentive to study kinetic traffic flow models in their own right.

As already observed in [66] certain kinetic models have the ability to produce a multivalued fundamental diagram. For a simplified Fokker–Planck type equation with this property we conduct a stability analysis in section 4.2. We obtain a result which seems surprising at first glance: for low and high density traffic flow is stable; at medium densities traffic behaves instable such that small changes to the density can produce stop–and–go waves. This corresponds to a behavior present in real–life scenarios.

4.1 Derivation of the AR–model from a Fokker–Planck type equation

This section contains an important result that demonstrates that kinetic models are a valid approach to model traffic flow on a single road. We show how the well–known and celebrated Aw–Rascle model can be derived from a kinetic equation [65].

The simplified model to be presented below relies on a model describing multilane traffic on a single road in terms of distribution functions [66] and we deduce the Aw–Rascle model from the former. In the following we briefly recall the original model and then introduce the simplified model which serves as a starting point for the derivation of the second–order partial differential equations model for traffic flow.

A Fokker–Planck type equation modeling traffic flow

We are concerned with Fokker–Planck type kinetic models for multi–lane traffic flow on a highway as introduced in [66]. The full Fokker–Planck model, homogenized over all lanes, is given by

$$\partial_t f + v\partial_x f + \partial_v \left(B[f](\rho, u, v - u)f - D_\epsilon[f](\rho, u, v - u)\partial_v f \right) = 0 \tag{4.1}$$

with boundary conditions at $v = 0$ and $v = v_{max}$ given by

$$B[f](\rho, u, v - u)f - D_\epsilon[f](\rho, u, v - u)\partial_v f = 0 \tag{4.2}$$

By $f = f(t, x, v)$ we denote the number density of cars which at time t are at location $x \in \mathbb{R}$ and move with speed $v \in [0, v_{max}]$. Here, v_{max} is the speed limit. $B[f]$ denotes a braking/acceleration force and $D_\epsilon[f]$ is a nondegenerate diffusive term modeling the inability of a driver to observe speeds with accuracy. For $\epsilon = 0, D_\epsilon = D_0$ may be degenerate, as used in [66].

In the original model [66] it was further assumed that all drivers have the same constant reaction time $\tau > 0$ and observe braking and acceleration thresholds $H_B := x + H_0 + T_B v$ and $H_A := x + H_0 + T_A v$, where v is the driver's speed and T_B and T_A are reaction times (in general different from τ). This means that a driver at x moving with speed v will brake in reaction to a traffic condition observed at $x + H_0 + T_B v$, and if no condition for braking applies, he/she will accelerate (if possible) in reaction to a traffic condition observed at $x + H_0 + T_A v$. As we will see below, the nonlocalities prove to be a key for deriving the AR–model from a simplified version of (4.1). H_0 is a safety distance in the range of one to three car lengths.

To model driver behavior, the braking term B and the diffusion coefficient D_ϵ in [66] were chosen as follows:

$$B[f](\rho, u, v - u) = \begin{cases} -c_B \rho (v - u^B)^2 (1 - P(u^B, v - u^B; b)) & v > u^B \\ c_A(\rho_{max} - \rho^A)(v - u^A)^2 & v \leq u^A \text{ and } v \leq u^B \\ 0 & \text{otherwise} \end{cases}$$

$$D_\epsilon[f](\rho, u, v - u) = \begin{cases} \sigma(\rho^B, u^B)|v - u^B|^\gamma + \epsilon & v > u^B \\ \sigma(\rho^A, u^A)|v - u^A|^\gamma + \epsilon & \text{otherwise} \end{cases}$$

Here, ρ^X and u^X for $X \in \{A, B\}$ incorporate the nonlocalities:

$$\rho^X := \rho(t - \tau, x + H_0 + T_X v) \tag{4.3a}$$

$$u^X := u(t - \tau, x + H_0 + T_X v) \tag{4.3b}$$

This exact form of the braking and diffusion term was chosen to make c_A and c_B dimensionless, cf. [66]. More importantly, for $v > u^B$ we decelerate and for $v < u^B$ we accelerate (if possible). $\sigma(\rho, u)$ is chosen in such a way that realistic values emerge at the endpoints of the fundamental diagram — we refer to [66] for details. The lane-changing probability was chosen as

$$P(u, v - u; b) = \begin{cases} b \cdot \left(\frac{v - u}{v_{max} - u}\right)^{\delta} & v > u \\ 0 & v \leq u \end{cases}$$

$\delta > 0$ is a parameter and $b \in [0, 1]$ can be used to model passing restrictions.

In the sequel we will consider a simplified Fokker–Planck type model with Robin boundary conditions, which retains the general features from (4.1). Henceforth we set the reaction time $\tau = 0$ and consider

$$\partial_t f + v \partial_x f + \partial_v \left(-bg(\rho^X)(v - u^X)f - d\partial_v f\right) = 0 \tag{4.4a}$$

$$-bg(\rho^X)(v - u^X)f - d\partial_v f = 0 \qquad v \in \{0, v_{max}\} \tag{4.4b}$$

Here, the index X again indicates nonlocal dependence of the dependent variables ρ, u (cf. (4.3)), and $g : \mathbb{R} \to \mathbb{R}^+$ is a sufficiently regular function, e.g., $g \in C^2(\mathbb{R})$. This version of the Fokker-Planck model is quite similar to the original model from [66] which we presented in (4.1). However, the original model includes a more complicated, non-factorized dependence on ρ and $v - u$, and here we use $(v - u^X)$ instead of $(v - u^X)|v - u^X|$, but still obtain the (crucial) correct qualitative behavior in the braking/acceleration term: for $v > u^X$ we decelerate whereas for $v < u^X$ we accelerate. In addition we take constant diffusion $d > 0$ and set $P(\ldots) = 0$. Finally, for simplicity we assume that the braking and acceleration delay is the same, i.e., we set $u^X = u^B = u^A$. Then the integrals to be evaluated during the derivation can easily be expressed in terms of the macroscopic quantities (4.5). In the original model we would distinguish $u^X = u^A$ and $u^X = u^B$ in the braking/acceleration terms, and that would produce more complicated expressions.

Our observation is that the simple choice $u^X = u^B$ leads to the AR–model [7].

From a simplified Fokker–Planck type equation to the AR–model

In this subsection we derive the well–known Aw–Rascle model [7] from the simplified Fokker–Planck equation presented above. As stated, some of the principal features of the more sophisticated model from [50, 66] are retained.

We provide a detailed description of our procedure, including the various simplifications necessary to obtain the AR–equations from our model equation. Not surprisingly, we have to neglect a number of terms arising from the Fokker–Planck formulation. In this sense the Aw–Rascle model emerges as a simplified description of traffic flow.

There are standard procedures to derive macroscopic (fluid–dynamic) models from kinetic equations [84]; of necessity, these procedures involve closure procedures for moment equations leading to different types of fluid dynamic descriptions, and the outcome depends on the scalings of independent and dependent variables as well as on the chosen closure process. In the context of traffic flow we mention [74].

In the present context we use only the zeroth and first moments. The macroscopic density ρ and flux $j = \rho u$, are given by

$$\rho \quad := \quad \int_0^{v_{max}} f(v)\, dv \tag{4.5a}$$

$$\rho u \quad := \quad \int_0^{v_{max}} v f(v)\, dv \tag{4.5b}$$

In view of (4.5a), (4.5b) is equivalent to

$$\int_0^{v_{max}} (v - u) f(v)\, dv \quad = \quad 0 \tag{4.6}$$

Before proceeding we recall one of the several versions in which the AR-model can be stated:

$$\partial_t \rho + \partial_x(\rho u) \quad = \quad 0 \tag{4.7a}$$
$$\partial_t(\rho u) + (-\rho p'(\rho) + u)\partial_x(\rho u) + (\rho u)\partial_x(u + p(\rho)) \quad = \quad 0 \tag{4.7b}$$

The form of the momentum equation (4.7b) is unconventional but is easily derived from the more common formulation

$$\partial_t(\rho(u + p(\rho))) + \partial_x(\rho u(u + p(\rho))) \quad = \quad 0.$$

We will derive equations (4.7) as moments of the simpler Fokker-Planck type kinetic equation (4.4)

$$\partial_t f + v \partial_x f + \partial_v \left(-bg(\rho^X)(v - u^X)f - d\partial_v f\right) = 0$$
$$-bg(\rho^X)(v - u^X)f - d\partial_v f = 0 \qquad v \in \{0, v_{max}\}$$

Consistent with the notation introduced above and from [66] we expand

$$\rho^X \quad := \quad \rho(t, x + H_X + T_X v) \approx \rho + (H_X + T_X v)\rho_x \tag{4.9a}$$
$$u^X \quad := \quad u(t, x + H_X + T_X v) \approx u + (H_X + T_X v)u_x \tag{4.9b}$$

where we have used a Taylor expansion up to first order. H_X is the typical car length (or minimal safety distance) and $T_X v$ is the distance a driver looks ahead in preparation for braking or other reactions to anticipated or observed traffic situations.

The no-flux boundary conditions (4.4b) implies the continuity equation (4.7a) from (4.4a) by a simple integration from 0 to v_{max}.

The equation for the first moment is more interesting. After multiplication of (4.4a) by v and integration by parts, using the boundary conditions (4.4b), we find

$$\partial_t j + \partial_x S + \int_0^{v_{max}} bg(\rho^X)(v - u^X)f - d\partial_v f\, dv \quad = \quad 0$$

where S denotes the second moment of f.

For the diffusion term we immediately find

$$\int_0^{v_{max}} d\partial_v f(v)\, dv \;=\; df(v)\,\big|_{v=0}^{v=v_{max}} = d(f(v_{max}) - f(0)) := D \tag{4.10}$$

If the distribution function is negligibly small at 0 and v_{max} then D is negligibly small.

Consider next the braking/acceleration term. Assuming that the nonlocality is of the form (4.9b) we have up to first order

$$(v - u^X) \;=\; v - u - (H_X + T_X v)u_x = -(u + H_X u_x) + (1 - T_X u_x)v \tag{4.11}$$

We group the terms by powers of v, because we are interested in moments of the distribution function f to recover our macroscopic quantities (4.5).

Furthermore, we expand

$$
\begin{aligned}
g(\rho^X) \;&=\; g(\rho + (H_X + T_X v)\rho_x) \\
&\approx\; g(\rho) + g'(\rho)(H_X + T_X v)\rho_x + \frac{1}{2}g''(\rho)(H_X + T_X v)^2\rho_x^2
\end{aligned} \tag{4.12}
$$

With this Taylor approximation the integral involving the braking term can be written as

$$\int_0^{v_{max}} b g(\rho^X)(v - u^X) f\, dv$$

$$\approx\; \int_0^{v_{max}} b\Big(g(\rho) + g'(\rho)H_X\rho_x + \frac{1}{2}g''(\rho)H_X^2\rho_x^2\Big)(v - u^X) f\, dv \tag{4.13a}$$

$$+ \int_0^{v_{max}} b\big(g'(\rho)T_X\rho_x + g''(\rho)H_X T_X\rho_x^2\big)v(v - u^X) f\, dv \tag{4.13b}$$

$$+ \frac{1}{2}\int_0^{v_{max}} b g''(\rho)T_X^2\rho_x^2 v^2(v - u^X) f\, dv \tag{4.13c}$$

In the sequel we will neglect the terms involving $g''(\rho)$ (this is, of course, exact if g is linear). To simplify (4.13), relation (4.6) plays a crucial role. Note that in general

$$0 \;=\; \int_0^{v_{max}} (v - u)f(v)\, dv \neq \int_0^{v_{max}} b g(\rho^X)(v - u)f(v)\, dv \tag{4.14}$$

because, by (4.12), $g(\rho^X)$ depends implicitly on v. We can incorporate this dependence by a Taylor expansion method but have to add correction terms in (4.14) involving higher moments of the distribution function f. However, we can use the relation

$$0 \;=\; \int_0^{v_{max}} (v - u)f(v)\, dv = \int_0^{v_{max}} b\big(g(\rho) + g'(\rho)H_X\rho_x\big)(v - u)f(v)\, dv \tag{4.15}$$

By using (4.15) in (4.13) we are led to

$$
\begin{aligned}
\int_0^{v_{max}} b g(\rho^X)(v - u^X) f\, dv \;&=\; \int_0^{v_{max}} -b\big(g(\rho) + g'(\rho)H_X\rho_x\big)(u^X - u) f\, dv \\
&\quad + \int_0^{v_{max}} b g'(\rho)T_X\rho_x v(v - u^X) f\, dv \\
&=\; \int_0^{v_{max}} -b\big(g(\rho) + g'(\rho)H_X\rho_x\big)(H_X u_x + T_X u_x v) f\, dv \\
&\quad + \int_0^{v_{max}} b g'(\rho)T_X\rho_x v(v - u^X) f\, dv
\end{aligned}
$$

Note that in the leading term on the right the dependence on v remains only through u^X. This is the key step in the derivation!

To proceed we consider first a simplified case: we neglect the terms involving $g'(\rho)$. Then we find the macroscopic equation

$$\partial_t(\rho u) + \partial_x S - bg(\rho)H_X u_x \rho - bg(\rho)T_X u_x \rho u - D = 0 \qquad (4.16)$$

We keep the diffusion induced term (4.10) on the right hand side. Furthermore, we can write

$$S = \int_0^{v_{max}} v^2 f\, dv = \int_0^{v_{max}} (v-u)^2 f\, dv - u^2 \rho + 2\rho u^2$$

Equation (4.16) becomes

$$\partial_t(\rho u) + \partial_x \left(\int_0^{v_{max}} (v-u)^2 f\, dv + u^2 \rho \right) - bg(\rho)H_X u_x \rho = D + b\frac{g(\rho)}{\rho} T_X \rho u_x j$$

Using $u_x \rho = (\rho u)_x - u\rho_x$ we find

$$\partial_t(\rho u) + \partial_x \left(\int_0^{v_{max}} (v-u)^2 f\, dv + u^2 \rho \right) - bg(\rho)H_X j_x + jb\frac{g(\rho)}{\rho}H_X \rho_x$$

$$= D + \frac{1}{2}b\frac{g(\rho)}{\rho}T_X (j^2)_x - jb\frac{g(\rho)}{\rho}T_X u\rho_x$$

Performing some of the differentiations and rearranging terms we are led to

$$\partial_t j + \left(-bg(\rho)(H_X + T_X u) + u \right) j_x + j\left(b\frac{g(\rho)}{\rho}(H_X + T_X u)\rho_x + u_x \right) \qquad (4.17a)$$

$$= -\partial_x \int_0^{v_{max}} (v-u)^2 f\, dv + D \qquad (4.17b)$$

The left-hand side here will take the same appearance as (4.7b) if we choose

$$p'(\rho) = b\frac{g(\rho)}{\rho}(H_X + T_X u).$$

Recall that up to first order the nonlocality was of the form $H_X + T_X v$. The above equation for $p'(\rho)$ also contains this expression, but the independent speed v has been replaced with the *average* speed u.

If we choose $g(\rho) = \rho$, then we obtain for the traffic–pressure $p(\rho) = b(H_X + T_X u)\rho \sim \rho$. We note that this satisfies the original conditions on $p(\rho)$ stated in [7] with $\gamma = 1$, provided $b, H_X, T_X > 0$.

Comparing (4.17) to (4.7b) we see that the only distinction is the source–term on the right hand side in (4.17). Without the partial derivative w.r.t. x this is the variance of the distribution about the mean velocity u. However, at present we are not able to estimate this term or argue that it will always be small.

Remark 4.1.1. *In our derivation we have neglected terms involving $g'(\rho)$. These terms can be retained, and computations similar to the ones presented above show that an Aw–Rascle model will emerge if we choose*

$$p'(\rho, \rho_x) \;=\; b\left(\frac{g(\rho)}{\rho}(H_X + T_X u) + \frac{g'(\rho)}{\rho}(H_X + T_X u)^2 \rho_x\right)$$

in order to obtain an equation similar to (4.7b). This result indicates that a higher order approximation in terms of the nonlocalities corresponds to a "higher order" approximation in p.

4.2 Stability Analysis of a Kinetic Equation for Traffic Flow

In this section we consider the stationary and spatially homogeneous version of the Fokker–Planck equation introduced in section 4.1. Using the boundary conditions (4.2) equation (4.1) reduces to

$$B[f](\rho, u, v - u)f \;=\; D_\epsilon[f](\rho, u, v - u)\partial_v f$$

In subsection 4.2.1 we will compute a stationary solution and in subsection 4.2.2 we will analyze the stability of this solution. We conclude our analysis with some numerical results in subsection 4.2.3.

In the sequel the zeroth and first moments of a stationary distribution function f will occur frequently, cf. section 4.1. The macroscopic density ρ and flux $j = \rho u$, are given by (cf. 4.5)

$$\rho \;:=\; \int_0^{v_{max}} f(v)\,dv \tag{4.18a}$$

$$\rho u \;:=\; \int_0^{v_{max}} v f(v)\,dv \tag{4.18b}$$

In view of relation (4.18a), (4.18b) is equivalent to

$$\int_0^{v_{max}} (v - u)f(v)\,dv \;=\; 0$$

4.2.1 Computation of a stationary distribution and the fundamental diagram

In this subsection we compute a stationary distribution for the Fokker–Planck equation for the nondegenerate case in which we replace the diffusion coefficient D with $D_\epsilon = D + \epsilon$. As we will see, the formulas become more complicated and –in contrast to the case discussed in [66]– the integration can not be perforemd analytically any more. However, we can make use of numerical techniques to compute the stationary distribution function. The stationary distribution function is a major ingredient in the computation in the fundamental diagram. The latter will play a vital role in the more elaborate algorithm used in the stability analysis.

Computation of a stationary and spatially homogeneous distribution function

We investigate the stationary and spatially homogeneous equation

$$B[f](\rho, u, v - u)f = D_\epsilon[f](\rho, u, v - u)\partial_v f \tag{4.19}$$

The braking term B and the diffusion coefficient D_ε are defined as follows

$$B[f](\rho, u, v - u) = \begin{cases} -c_B \rho (v-u)^2 (1 - P(u, v-u; b)) & v > u \\ c_A(\rho_{max} - \rho)(v-u)^2 & v \leq u \end{cases} \tag{4.20}$$

$$D_\epsilon[f](\rho, u, v - u) = \begin{cases} \sigma(\rho, u)|v-u|^\gamma + \epsilon & v > u \\ \sigma(\rho, u)|v-u|^\gamma + \epsilon & v \leq u \end{cases} \tag{4.21}$$

$\sigma(\rho, u)$ is chosen as in [66]. For the lane-changing probability we choose

$$P(u, v-u; b) = \begin{cases} b \cdot \left(\frac{v-u}{v_{max}-u}\right)^\delta & v > u \\ 0 & v \leq u \end{cases} \tag{4.22}$$

Note that this is similar to the choice in [66] except that we explicitly introduce a 'residual' diffusion ϵ to get rid of the degeneracy and we have an additional control parameter b in the lane–change probability.

Due to the structure of B and D_ϵ given by (4.20) and (4.21), respectively, we need to solve different ordinary differential equations (ODEs) in (4.19) in the two cases $v > u$ and $v < u$. Note that we are interested in a continuous stationary distribution which will eliminate one degree of freedom in the integration constants. Additionally, we need to ensure that ρ and f are related via (4.5); this fully determines the integration constants. To keep the formulas tractable, we will choose $\delta = \gamma = 1$ throughout the remainder of this work.

Remark 4.1. *For $\gamma \neq 1, \delta \neq 1$ we can still find closed expressions for the primitives of the ODE under consideration but they are lengthy. We need to compute an integral of the form ($p = 2$ and $p = 2 + \gamma$, $q = \delta$, $x = v - u$)*

$$\int \frac{x^p}{ax^q + b} \, dx \tag{4.23}$$

In principle one could try a substitution of the form $z = ax^q + b$ or $x = \left(\frac{z-b}{a}\right)^{\frac{1}{q}}$ and then end up with an integral of the form

$$\frac{1}{q} \int \frac{1}{z} \frac{1}{z-b} \left(\frac{z-b}{a}\right)^{\frac{p+1}{q}} dz = \frac{1}{q} a^{-\frac{p+1}{q}} \int \frac{1}{z} (z-b)^{\frac{p-q+1}{q}} dz \tag{4.24}$$

Now one can use the generalized binomial series and obtain an expression in powers of z that can be 'easily' integrated.

Assume we are given a pair (ρ, u). Then we can solve the ODE (4.19) numerically. For $v < u$ the solution to the ODE takes the form

$$f(v) = e^{c_1(\rho, u)} \cdot e^{G_1(v, \rho, u)} \tag{4.25}$$

106

and for $v > u$ we obtain

$$f(v) \;\; = \;\; e^{c_2(\rho,u)} \cdot e^{G_2(v,\rho,u)+G_3(v,\rho,u)} \tag{4.26}$$

The exact expressions for G_1, G_2 and G_3 can be found in (4.33). Due to our continuity constraint we require that $\lim_{v \uparrow u} f(v) = \lim_{v \downarrow u} f(v)$. Using (4.25) and (4.26) we can compute these limits and find

$$c_1 \;\; = \;\; M + c_2 \tag{4.27}$$

with an $M = M(\rho, u, b, v_{max}, \rho_{max}, \sigma, \epsilon, c_A, c_B)$; the exact form is stated in (4.34) below.

The stationary distribution function has to satisfy the additional constraint

$$\rho \;\; := \;\; e^{c_2} \left(e^M \int_0^u e^{G_1(v)} \, dv + \int_u^{v_{max}} e^{G_2(v)+G_3(v)} \, dv \right) \tag{4.28}$$

Analytically this formula is perfectly fine. It turns out that to avoid numerical difficulties in the evaluation of the integrals in (4.28) we should introduce a parameter c_3. We choose it as $c_3 := M + \max_{v<u} G_1(v) = M + G_1(u)$ and rewrite (4.28)

$$\rho \;\; := \;\; e^{c_2+c_3} \left(\int_0^u e^{G_1(v)-G_1(u)} \, dv + e^{-M-G_1(u)} \int_u^{v_{max}} e^{G_2(v)+G_3(v)} \, dv \right) \tag{4.29}$$

This determines the factor $e^{c_2+c_3}$ uniquely.

More explicitly, we can evaluate the quantities

$$M_1 := \int_0^u e^{G_1(v)+M-c_3} \, dv \qquad M_2 := \int_u^{v_{max}} e^{G_2(v)+G_3(v)-c_3} \, dv \tag{4.30}$$

using a numerical quadature routine and then compute

$$e^{c_2+c_3} \;\; = \;\; \frac{\rho}{M_1 + M_2} \tag{4.31}$$

to scale the distribution function to the correct value. Then e^{c_2} and $e^{c_1} = e^M e^{c_2}$ are determined.

Finally we can write down an expression for the stationary distribution function:

$$f(v; \rho, u, \epsilon) \;\; = \;\; \begin{cases} \dfrac{\rho}{M_1+M_2} \cdot e^{G_1(v,\rho,u,\epsilon)-G_1(u,\rho,u,\epsilon)} & v \leq u \\[2mm] \dfrac{\rho}{M_1+M_2} \cdot e^{G_2(v,\rho,u,\epsilon)+G_3(v,\rho,u,\epsilon)-M-G_1(u,\rho,u,\epsilon)} & v \geq u \end{cases} \tag{4.32}$$

As $\epsilon \to 0$ we obtain –at least formally– the same stationary distribution as in [66].

For sake of completeness we give the expressions for the functions $G_i, i = 1, 2, 3$ and

107

M. They read

$$G_1(v) \; := \; c_A(\rho_{max} - \rho) \left(-\frac{(-\sigma(v-u) + \epsilon)^2}{2\sigma^3} + \frac{2\epsilon(-\sigma(v-u) + \epsilon)}{\sigma^3} \right.$$
$$\left. -\frac{\epsilon^2}{\sigma^3} \ln(-\sigma(v-u) + \epsilon) \right) \tag{4.33a}$$

$$G_2(v) \; := \; -c_B\rho \left(\frac{(\sigma(v-u) + \epsilon)^2}{2\sigma^3} - \frac{2\epsilon(\sigma(v-u) + \epsilon)}{\sigma^3} \right.$$
$$\left. +\frac{\epsilon^2}{\sigma^3} \ln(\sigma(v-u) + \epsilon) \right) \tag{4.33b}$$

$$G_3(v) \; := \; b \cdot \frac{c_B\rho}{v_{max} - u} \cdot \left(\frac{(\sigma(v-u) + \epsilon)^3}{3\sigma^4} - \frac{3\epsilon(\sigma(v-u) + \epsilon)^2}{2\sigma^4} \right. \tag{4.33c}$$
$$\left. +\frac{3\epsilon^2(\sigma(v-u) + \epsilon)}{\sigma^4} - \frac{\epsilon^3}{\sigma^4} \ln(\sigma(v-u) + \epsilon) \right) \tag{4.33d}$$

For M we have

$$M \; := \; \frac{\epsilon^2}{\sigma^3} c_A(\rho_{max} - \rho) \left(-\frac{3}{2} + \ln(\epsilon) \right) \tag{4.34a}$$
$$-\frac{\epsilon^2}{\sigma^3} c_B\rho \left(-\frac{3}{2} + \ln(\epsilon) + \frac{b}{v_{max} - u} \frac{\epsilon}{\sigma} \cdot \left(-\frac{11}{6} + \ln(\epsilon) \right) \right) \tag{4.34b}$$

Additional precautions for small values of σ

If we choose $\sigma = const. \ll 1$ our formulas (4.33) used to compute a stationary distribution are numerically not stable. In particular, for $\sigma = 0$ the functions G_i, $i = 1, 2, 3$ are not defined. The problem can be tackled by solving two different sets of ODEs depending on the conditions $v < u$ and $v > u$. We consider the case $\sigma = 0$ in the following.

For $v > u$ we need to solve the ODE

$$\partial_v f \; = \; \left(-c_B\rho \frac{(v-u)^2}{\epsilon} + b \frac{c_B\rho}{(v_{max} - u)^\delta} \frac{(v-u)^{2+\delta}}{\epsilon} \right) f \tag{4.35}$$

As before we choose $\delta = 1$. Furthermore, we can solve the ODE easily. We obtain

$$f(v) \; = \; c_1 e^{G_1(v)}, \qquad v > u \tag{4.36a}$$
$$G_1(v) \; = \; -\frac{1}{3\epsilon} c_B\rho(v-u)^3 + \frac{1}{4\epsilon} \frac{c_B\rho}{(v_{max} - u)} (v-u)^4 \tag{4.36b}$$

Now we consider $v < u$; in this case we need to solve

$$\partial_v f = \frac{c_A(\rho_{max} - \rho)(v-u)^2}{\epsilon} f$$

In this case the solution is

$$f(v) = c_2 e^{G_3(v)}, \qquad v < u \tag{4.37a}$$

$$G_3(v) = \frac{1}{3\epsilon} c_A (\rho_{max} - \rho)(v - u)^3 \tag{4.37b}$$

As one observes, for continuity we need $c_1 = c_2$. From the condition

$$\rho = \int_0^{v_{max}} f(v) \, dv$$

we then obtain

$$c_1 = c_2 = \rho \left(\int_0^u e^{G_3(v)} \, dv + \int_u^{v_{max}} e^{G_1(v)} \, dv \right)^{-1}$$

which is similar to our previous formulas. The appearing integrals can be evaluated using a numerical integration scheme. These formulas enable us to compute the correct stationary distribution for small values $\sigma = const \le 10^{-4}$. We set $\sigma = 0$ in our computations in this case and use the formulas (4.36) and (4.37).

Computation of the fundamental diagram

Above we presented a method to compute a solution to the stationary and spatially homogeneous equation (4.19). This enables us to compute the so–called fundametal diagram. The latter is important for traffic engineers to describe traffic flow. The fundamental diagram F is the set

$$\begin{aligned}
F &:= \left\{ (\rho, u) \mid \rho u = j = \int_0^{v_{max}} v f(v) \, dv \right\} \\
&= \left\{ (\rho, u) \mid \int_0^{v_{max}} (v - u) f(v) \, dv = 0 \right\} \tag{4.38}
\end{aligned}$$

The results presented so far enable us to compute a distribution function f for a given pair (ρ, u) and a value of ϵ satisfying

$$\rho = \int_0^{v_{max}} f(v) \, dv$$

As described below we can, for example, use a bisection technique in order to compute the fundamental diagram F. This is not new and results on this can be found in [50, 66]. In particular, one obtains multivalued fundamental diagrams in these computations for certain parameter sets.

We start out with a given u and three values $\rho_l < \rho_m < \rho_r$. Then we compute $R_l := R(\rho_l; u)$, $R_m := R(\rho_m; u)$ and $R_r := R(\rho_r; u)$ where

$$R(\rho; u) = \int_0^{v_{max}} (v - u) f(v) \, dv$$

f is the stationary distribution that is computed for a given pair of (ρ, u) as described in the previous section. If $\frac{R_l}{R_r} > 0$ the bisection method to find a zero of $R(\rho; u)$ will not work. In this case we need to choose ρ_l and ρ_r differently. Otherwise we update one of the values of ρ_l and ρ_r with ρ_m, compute a new ρ_m and iterate.

In previous works it was shown that for $\epsilon = 0$ and for a given u there is at least one ρ^* s.t. $R(\rho^*; u) = 0$. In essence, the argument is a simple application of the intermediate value theorem. In the present situation it is not as easy to estimate the integrands (given by (4.32)) and show that $R(\rho)$ always needs to posess at least one zero for a given u and ϵ. Nevertheless, the outlined procedure works well.

4.2.2 Stability of spatially homogeneous distributions

In this subsection we investigate the spatially homogeneous equation

$$\partial_t f + \partial_v(B[f](\rho, u, v - u)f - D_\epsilon[f](\rho, u, v - u)\partial_v f) = 0 \qquad (4.39)$$
$$B[f](\rho, u, v - u)f - D_\epsilon[f](\rho, u, v - u)\partial_v f = 0 \qquad v \in \{0, v_{max}\}$$

We will derive the linearization around a stationary distribution f_0 in direction $\tilde{f}$ for equation (4.39). We set $f = f_0 + \tilde{f}$. Then we use a Taylor expansion and collect the linear terms in $\tilde{f}$. Note that f_0 solves the stationary equation and therefore the expression should simplify. For the perturbation we will use an ansatz as in [94]:

$$\tilde{f}(t, v) = r(t)e^{ikv + \lambda t} << 1 \qquad (4.40)$$

k is the wave number and we want to determine $\lambda = \lambda(k)$ as function of k. The linearization will lead to a formula for λ for which we can then conduct a bifurcation analysis, see below. In the context of traffic flow a bifurcation analysis for a different traffic model can be found in [36].

In the sequel we set set $\tilde{\rho} := \rho_f(f_0)[\tilde{f}]$ and $\tilde{u} := u_f(f_0)[\tilde{f}]$. A linearization for a general perturbation $\tilde{f}$ of (4.39) is given by

$$0 = \partial_t \tilde{f} + \partial_v((B_\rho \cdot \tilde{\rho} + B_u \cdot \tilde{u})f_0 \qquad (4.41a)$$
$$+ (B + B_\rho \cdot \tilde{\rho} + B_u \cdot \tilde{u})\tilde{f} \qquad (4.41b)$$
$$- (D_\epsilon)_u \tilde{u}\partial_v f_0 - (D_\epsilon + (D_\epsilon)_u \tilde{u})\partial_v \tilde{f}) \qquad (4.41c)$$

where B_ρ, B_u and $(D_\epsilon)_u$ are evaluated at $(\rho, u) = (\rho(f_0), u(f_0))$. The boundary conditions at $v = 0$ and $v = v_{max}$ read

$$0 = (B_\rho \cdot \tilde{\rho} + B_u \cdot \tilde{u})f_0 \qquad (4.42a)$$
$$+ (B + B_\rho \cdot \tilde{\rho} + B_u \cdot \tilde{u})\tilde{f} \qquad (4.42b)$$
$$- (D_\epsilon)_u \tilde{u}\partial_v f_0 - (D_\epsilon + (D_\epsilon)_u \tilde{u})\partial_v \tilde{f} \qquad (4.42c)$$

Note that if we integrate the linearized equation (4.41) with linearized boundary conditions (4.42) for a perturbation $\tilde{f}$, we are led to

$$\partial_t \int_0^{v_{max}} \tilde{f} \, dv = 0$$

With our ansatz (4.40) $\tilde{f}(t,v) = r(t)e^{ikv+\lambda t}$ we then obtain

$$
\begin{aligned}
0 &= \partial_t \int_0^{v_{max}} r(t)e^{ikv+\lambda t}\,dv = \partial_t(r(t)e^{\lambda t}\frac{1}{ik}(e^{ikv_{max}} - 1)) \\
&= \frac{1}{ik}(e^{ikv_{max}} - 1)\left(r'(t)e^{\lambda t} + r(t)\lambda e^{\lambda t}\right)
\end{aligned}
\tag{4.43}
$$

Of course, we have $e^{\lambda t} \neq 0$ and therefore we obtain two equations

$$
\begin{aligned}
r'(t) &= -\lambda r(t) & \text{(4.44a)} \\
e^{ikv_{max}} - 1 &= 0 & \text{(4.44b)}
\end{aligned}
$$

(4.44a) leads to $r(t) = c \cdot e^{-\lambda t}$ for some $c \in \mathbb{R}$. Then our ansatz (4.40) leads to a stationary perturbation which we do not want to consider. Therefore, we require

$$
k = 2\pi l \frac{1}{v_{max}}, \quad l \in \mathbb{Z}
\tag{4.45}
$$

At this point we need to compute the quantities $\tilde{\rho}$ and $\tilde{u}$, since they contain $\tilde{f}$. In order to obtain the expressions, we need to compute the integrals which define them. We have for $\tilde{\rho}$

$$
\begin{aligned}
\tilde{\rho} = \rho_f(f_0)[\tilde{f}] = \int_0^{v_{max}} \tilde{f}\,dv &= r(t)e^{\lambda t}\frac{1}{ik}e^{ikv}\,|_0^{v_{max}} \\
&= -\frac{1}{k^2}r(t)e^{\lambda t}(e^{ikv_{max}} - 1)i \\
&= -\frac{1}{k^2}(e^{ikv_{max}} - 1)ie^{-ikv}\tilde{f} \\
&= 0
\end{aligned}
$$

where the last equality holds due to (4.44b). For $\tilde{u}$ we are led to

$$
\begin{aligned}
\tilde{u} = u_f(f_0)[\tilde{f}] = \frac{1}{\rho}\int_0^{v_{max}} v\tilde{f}\,dv &= \frac{1}{\rho}r(t)e^{\lambda t}\int_0^{v_{max}} ve^{ikv}\,dv \\
&= \frac{1}{\rho}r(t)e^{\lambda t}\left(\frac{v}{ik}e^{ikv}\,|_0^{v_{max}} - \int_0^{v_{max}} e^{ikv}\,dv\right) \\
&= \frac{1}{\rho}r(t)e^{\lambda t}\left(-\frac{1}{k^2} + e^{ikv_{max}}(-i\frac{1}{k} + \frac{1}{k^2})\right) \\
&= \frac{1}{\rho}\left(-\frac{1}{k^2} + e^{ikv_{max}}(-i\frac{1}{k} + \frac{1}{k^2})\right)e^{-ikv}\tilde{f} \\
&= \frac{1}{\rho}\frac{1}{ik}e^{-ikv}\tilde{f}
\end{aligned}
$$

where the last equality hold again because of our special requirement on k, cf. (4.45). Note that $\tilde{u}$ is linear in $\tilde{f}$ and therefore some of the terms appearing in the general linerization (4.41) can be neglected in a linearization. Furthermore, we obtain for our special choice of the perturbation $\tilde{f}$

$$
\begin{aligned}
\partial_t \tilde{f} &= r'(t)e^{ikv+\lambda t} + \lambda \tilde{f} & \text{(4.46a)} \\
\partial_v \tilde{f} &= ik\tilde{f} & \text{(4.46b)}
\end{aligned}
$$

Exploiting the relations for $\tilde{\rho}$ and $\tilde{u}$ and (4.46), (4.41) reduces to

$$0 = \left(\frac{r'(t)}{r(t)} + \lambda\right)\tilde{f} + \partial_v\left((B_u \cdot \frac{1}{\rho}\frac{1}{ik}e^{-ikv}\tilde{f})f_0\right) \tag{4.47a}$$

$$+ B\tilde{f} - (D_\epsilon)_u \frac{1}{\rho}\frac{1}{ik}e^{-ikv}\tilde{f}\,\partial_v f_0 - ikD_\epsilon\tilde{f}\right) \tag{4.47b}$$

Let us define the linear operator (w.r.t. $\tilde{f}$) $M(v)\tilde{f}$ so that (4.47) can be written as

$$0 = \left(\frac{r'(t)}{r(t)} + \lambda\right)\tilde{f} + \partial_v(M(v)\tilde{f})$$

Then we are led to

$$0 = \left(\frac{r'(t)}{r(t)} + \lambda\right)\tilde{f} + (\partial_v M(v))\tilde{f} + M\partial_v\tilde{f}$$

$$= \left(\frac{r'(t)}{r(t)} + \lambda\right)\tilde{f} + (\partial_v M(v))\tilde{f} + ikM(v)\tilde{f}$$

Clearly, this equation has the structure

$$0 = (r'(t) + (\lambda - A)r(t))\tilde{f} \tag{4.48}$$

For our stability analysis, we are only interested in the eigenvalues of the operator A and consider therefore

$$0 = \lambda \cdot id - A$$

or in our particular case

$$0 = \lambda \cdot id - (-\partial_v M(v) - ikM(v)) \tag{4.49}$$

Since we still have a v–dependence, we integrate (4.49). Due to the boundary conditions for the linearized equation, we have

$$0 = \lambda v_{max} + ik\int_0^{v_{max}} M(v)\,dv \tag{4.50}$$

Therefore, we can compute λ in dependence of the wave number k and for a given pair (ρ, u) by

$$\lambda(k; \rho, u) = \frac{1}{v_{max}}\left(-ik\int_0^{v_{max}} M(v)\,dv\right)$$

$$= \frac{1}{v_{max}}\left(-ik\int_0^{v_{max}} \frac{1}{\rho}\frac{1}{ik}e^{-ikv} \cdot B_u \cdot f_0\right) \tag{4.51a}$$

$$+ B - \frac{1}{\rho}\frac{1}{ik}e^{-ikv}(D_\epsilon)_u\,\partial_v f_0 - ikD_\epsilon\,dv\right) \tag{4.51b}$$

For sake of completeness we state the expressions for the occuring partial derivatives. We have

$$B_u(\rho(f), u(f)) = \begin{cases} -c_B\rho(-2(v-u)(1-P(u,v-u;b)) \\ \qquad -(v-u)^2\partial_u P(u,v-u;b)) & v > u \\ \qquad -2c_A(\rho_{max} - \rho)(v-u) & v \leq u \end{cases}$$

112

and

$$(D_\epsilon)_u(\rho(f), u(f)) = \begin{cases} -\sigma\,\gamma(v-u)^{\gamma-1} & v > u \\ \sigma\,\gamma(u-v)^{\gamma-1} & v \leq u \end{cases}$$

4.2.3 Numerical Results

In a first test–setup, we choose $\sigma = 0$, $c_A = 5$, $c_B = 25$, $\rho_{max} = v_{max} = 1$, $\epsilon = 0,01$ and $b = 1$. For the computation of the fundamental diagram we discretize the interval $[0.16, 0.82]$ with the constant width 0.1. For every $u_j \in [0.16, 0.82]$ we then compute the corresponding ρ as outlined in subsection 4.2.1. We obtain a multi–valued fundamental diagram as in figure 4.1. As an example for a stationary distribution, we pick one pair $(\rho(u_j), u_i)$ and depict the corresponding distribution function in figure 4.2.

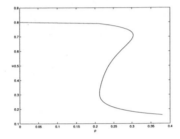

Figure 4.1: Multi–valued fundamental diagram.

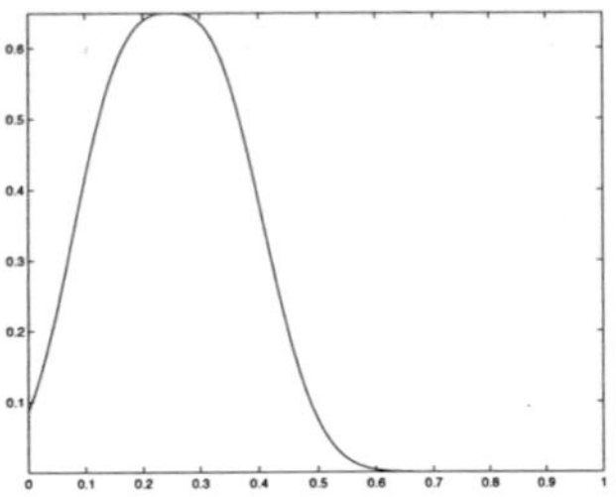

Figure 4.2: Sample stationary distribution function for $(\rho(u_j), u_j) = (0.2218, 0.25)$ on the fundamental diagram 4.1.

In figures 4.4 – 4.7 we show the eigenvalues corresponding to perturbations of the stationary distribution functions on the fundamental diagram from figure 4.1. The eigenvalues $\lambda(k)$, cf. (4.51), are depicted for $k = -10, \ldots, 10$. We note that the code is symmetric in the sense that $\lambda^*(k) = \lambda^*(-k)$. This matches the expectations one has from the underlying theory. If one looks at their behavior we realize that the zeroth mode $(k = 0)$ is stable and the instability comes from the first mode $(k = 1$ or $k = -1)$.

We observe that the perturbation is stable for small values of u. However, in the interval $u \in (0.35, 0.36)$ the real part of the maximal eigenvalue changes sign. In figure 4.4 the maximal value is $\lambda_{max} = -0.013068$ and in figure 4.5 we have $\lambda_{max} = 0.022002$. We obtain an instable perturbation in the region $u \in (0.36, 0.64)$ for this example, i.e., $\lambda_{max} > 0$ for this region. For $u = 0.65$ the sign changes once more and we recover a stable perturbation. In figure 4.6 the maximal value is $\lambda_{max} = 0.0042266$ and in figure 4.7 we have $\lambda_{max} = -0.022247$.

Furthermore, we observe that $\lambda^*(k)$ varies smoothly as function of u (note that for a given u a particular $\rho = \rho(u)$ is determined via the fundamental diagram).

For this example, we summarized our observations in figure 4.3.

113

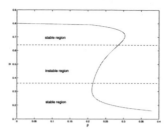

Figure 4.3: Stability regions.

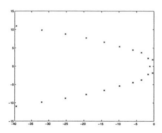

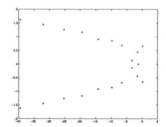

Figure 4.4: Eigenvalue $\lambda(k)$ from (4.51) for $u_j = 0.25$; here we still have stability.

Figure 4.5: Eigenvalue $\lambda(k)$ from (4.51) for $u_j = 0.39$; here we loose stability.

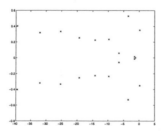

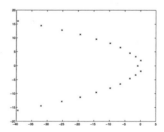

Figure 4.6: Eigenvalue $\lambda(k)$ from (4.51) for $u_j = 0.42$; here the perturbation still leads to instabilities.

Figure 4.7: Eigenvalue $\lambda(k)$ from (4.51) for $u_j = 0.62$; here we recover stability.

114

Finally, we study the behavior of the first mode along the fundamental diagram. More explicitly, we plot in figures 4.8 and 4.9 the real and imaginary part of $\lambda(1, \rho, u(\rho))$, respectively. Note that for certain values of ρ we have a triple–valued fundamental diagram, cf. figure 4.1, and therefore the curve for $\lambda(1; \rho, u(\rho))$ has three distinct branches which are visible. Furthermore, we see from figure 4.8 once more that two solutions are stable and one is instable. The second mode is displayed in figures 4.10 and 4.11 and the third one can be found in figures 4.12 and 4.13. Here we do not encounter instabilities.

At present, the interpretation of this behavior is not completely understood. We expected one stable point and two unstable ones in the multivalued region. However, the results do not leave room for this conclusion. However, in semiconductors there is a similar result [94]. The best explanation we were able to come up with reads as follows:

For low speeds and higher densities traffic flow is rather uniform and hence small deviations from the mean speed do not have a too severe consequence for the overall behavior. Therefore, the distribution is stable for smaller values of u. For large values of u and small densities we have the situation that there are not too many cars going at a high speed. Therefore, they do not interact too much. Then small deviations do not affect the distribution and we obtain a stable situation. However, for a region around $\frac{v_{max}}{2}$ (or for values of u that are not too extreme) with a sufficiently large density small perturbations play a role and cause traffic jams (which is in accordance with real life observations). This is reflected in the fact that we have an instable part of the fundamental diagram in the region $u \in (0.36, 0.64)$ in our example.

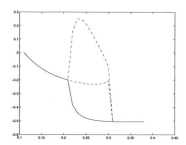

Figure 4.8: Real part of the first mode in dependence of $(\rho, u(\rho))$ corresponding to the fundamental diagram from figure 4.1.

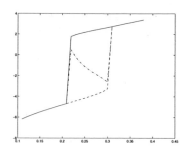

Figure 4.9: Imaginary part of the first mode in dependence of $(\rho, u(\rho))$ corresponding to the fundamental diagram from figure 4.1.

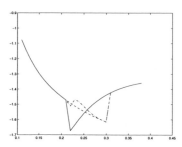

Figure 4.10: Real part of the second mode.

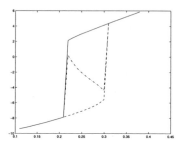

Figure 4.11: Imaginary part of the second mode.

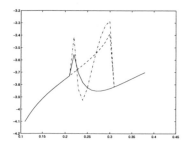

Figure 4.12: Real part of the third mode.

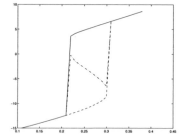

Figure 4.13: Imaginary part of the third mode.

116

Chapter 5

Conclusion and Outlook

The main results of this thesis can be summarized as follows:

- We devise a multi–class model on traffic flow networks that respects driver's destination preferences, cf. subsections 2.1.2, 2.1.3 and appendix A.

- We show the relationship of the Aw–Rascle traffic flow model to a kinetic equation, cf. section 4.1.

- We derive a framework for optimization on networks in which the dynamics is governed by scalar conservation laws, cf. Theorem 2.2.17.

In this thesis we were mainly concerned with macroscopic traffic flow models on networks. Additionally, we investigated related optimal control problems which belong to a more general framework.

Our result concerning the multi–class or multicommodity model is based on ideas and models presented in [19, 34] which were reviewed briefly in subsection 2.1.1. In the sequel we devised our multicommodity model. In particular we modified the coupling conditions. We show with selected examples how to use the formulas and how they relate to existing road network models [34, 51]. Furthermore, we discussed the solution structure of the new model in detail and show that we can always construct a weak solution. For the special case of two classes we conduct a more detailed analysis of how to solve Riemann–problems for junctions of a particular type in subsection 2.1.4.

We conclude the investigation of multi–class models in subsection 2.3.1. The presented numerical results indicate that the multi–commodity model is correct in the sense that the solution reflects driver's preferences: everyone arrives at their desired destination. It turns out that we can interpret the new coupling conditions as control constraints for the reference network model.

Another key result in this thesis can be found in section 4.1. We showed that a simplified kinetic model yields the macroscopic Aw–Rascle model. The derivation implies that kinetic traffic flow models can contain more information than macroscopic ones. Therefore, it is worth studying them in their own right. This remark motivates the stability analysis we include in section 4.2 for a particular kinetic traffic flow model. Additionally, it might we worthwile to investiagte kinetic models in a network setup. This presents new challenges in the way coupling conditions need to be designed. Furthermore, questions on suitable algorithms for reliable simulations need to be addressed.

The third aspect treated in this thesis concerns optimization. In subsection 2.2.1 we review a general framework for PDE–constrained optimization. We use it as a basis for the formal derivation of first order necessary optimality conditions for a traffic and supply network. Our presented formulas are not mathematically rigorous since we did not prove that all the appearing operators are well–defined in our setup. We merely assumed that this was the case. It is desirable to establish results which make the outlined procedure mathematically rigorous. This is beyond the scope of this thesis.

However, the numerical results subsections 2.2.2 and 3.2.2 indicate that the derived equations are correct. In particular, we mention that a comparison of the gradient computed with finite differences and our adjoint formulas leads to the same results, cf. subsection 3.2.1. The proposed optimization algorithm, essentially a projected gradient method, works reliably. The procedure becomes less efficient if we increase the number of time–intervals for the optimization or the network size.

In this thesis the largest network under consideration consists of 32 roads. Clearly, in practical applications one wants to consider larger networks. In principal, this is possible with the methods presented in this work and the code is also able to handle larger networks. However, the computation time increases drastically and in the end prohibits the usage of PDE–based adjoint calculus for the optimization.

Therefore, we investigate simplified models and related optimization problems. A particularly intriguing idea is the usage of receeding horizon controls: the optimization on the whole time–interval $[0, T]$ is done independently on a sequence of smaller time–intervals. In subsections 2.2.2 and 2.3.2 we presented an instantaneous control approach and corresponding numerical results. Our studies show that the method works well and reduces both computation time and storage requirements by a significant factor.

Other model simplifications are possible; for example one can derive simplified models via Front Tracking. This approach is currently under investigation. Other approaches are based on averaging the occuring quantities over single roads, cf. [48, 51, 75].

Ultimately we want to relate the multi–class models to mixed–integer linear programs (MIPs) so that discrete optimization techniques can be exploited. They work particulary well on large networks. Additionally, it would be beneficial to see how the optimal controls from the MIP and the adjoint–based controls relate for a multi–class traffic flow model.

As briefly highlighted in subsection 3.1.1 a model hierachy relating PDE models to MIPs exists for supply networks in a single–class setting. With the presented adjoint calculus we can compute optimal controls and compare them to the ones from the MIP. From an engineer's point of view the controls resulting from the adjoint approach are easier to implement in a machine. Additionally, we were able to verify that the two different approaches computed the same values for the objective functional under consideration. We hope that we can extend these results to multi–commodity supply networks to model applications from industry more closely.

As the numerical results in subsections 2.3.2 and 3.2.2 indicate, adjoint calculus is a powerful tool to compute optimal controls reliably. However, throughout this work we have made one crucial assumption: we do not consider state constraints, i.e., we only cover the case $Y_{ad} = Y$. For $Y_{ad} \subset Y$ there is a theory of how to derive first order necessary conditions (i.e., there is an analogue to corollary 2.2.9), but the optimality system becomes much more complicated even on a *single* road. To the best of our knowledge there are no results on any kind of network.

For practical applications, such as supply networks, restrictions on the state are present; for example, queues have a certain upper bound. Additional constraints arise from the particular objective under investigation. Processor shutdown due to maintenance falls into this category as well as the consideration of min–up or min–down times of certain processors or queues. This brief description indicates that an extension and generalization of the adjoint calculus framework to this more general situation is desirable. As mentioned before, this extension can be difficult to handle mathematically and analytically. From a practical point of view a possible solution is to apply automatic differentiation (AD) [42] to a forward solver for a particular model.

The design of suitable optimization routines for the more elaborate multicommodity model is a work in progress. At present, an adjoint–based strategy is under investigation. However, since the commodity equations are not in conservative form, the multicommodity model does not exactly fit into the framework of Theorem 2.2.17. Additionally, we envision usage of the technique from the instantaneous control approach in the derivation of more tractable optimization problems. Furthermore, we believe that AD can contribute to an efficient optimization routine for the complex multicommodity network model.

Appendix A

Detailed analysis of the solution structure for the commodity equations

In this appendix we will discuss the general solution structure of the multi–commodity equation. More precisely, we will be concerned with the solution of Riemann problems:

$$\partial_t \gamma^i(t,x) + v(t,x) \cdot \partial_x \gamma^i(t,x) \;=\; 0 \tag{A.1a}$$

$$\gamma^i(0,x) \;=\; \left\{ \begin{array}{ll} \gamma_l^i & x < 0 \\ \gamma_r^i & x > 0 \end{array} \right. \tag{A.1b}$$

where v is discontinuous and given by

$$v(t,x) \;=\; \left\{ \begin{array}{ll} v_l & x - st < x_0 \\ v_r & x - st > x_0 \end{array} \right. \tag{A.2}$$

We assume w.l.o.g. $x_0 > 0$ and we will supress the commodity index i in (A.1) in the sequel. Additionally, we note that v_l, v_r are independent of the initial data for the Riemann-problem for equation (A.1a). Thus, our case differs from the example (Burgers equation with a discontinuous function a [in our notation $a = v$]) presented in [93], section 5.

As we observed in subsection 2.1.3 the Filippov–characteristics [31] play a key role in the construction of the solution. Recall that in the LWR–model the velocity v and the density ρ are essentially related via $v(\rho) = c\,(1 - \rho)$ for some $c \in \mathbb{R}^+$. We will show that in our application the generalized characteristics are uniquely determined in either case $\rho_l < \rho_r$ or $\rho_l > \rho_r$, i.e. independent of $v_l > v_r$ or $v_l < v_r$. This ensures the solvability of (A.1).

Additionally, we will discuss procedures to define measure solutions in cases in which the Filippov characteristics are not unique. This can be done using the notion of *reversible* and *duality* solution introduced in [14].

Connection to the LWR model

In our multi–commodity model the velocity $v(x,t)$ is related to the solution of the LWR–equation (A.4a) via

$$v(\rho) \;=\; c\,(1-\rho), \qquad 0 \leq \rho \leq 1, \qquad c \in \mathbb{R}^+ \tag{A.3}$$

In the following we will use $v_l := v(\rho_l)$ and $v_r := v(\rho_r)$ to keep the notation reasonably short.

We need to consider Riemann-problems for the LWR–equation (A.4a)

$$\partial_t \rho + \partial_x f(\rho) \;=\; 0 \tag{A.4a}$$

$$\rho_0(x) \;=\; \begin{cases} \rho_l & x \leq x_0 \\ \rho_r & x > x_0 \end{cases} \tag{A.4b}$$

The correct entropy solution [77] for (A.4a) depends on the relation of ρ_l and ρ_r. Due to our assumptions (2.6), the flux function f is concave; a prototype for f is depicted in figure A.1. We distinguish two cases in the solution of (A.4)

1. $\rho_l < \rho_r$
   The entropic solution is a shock wave traveling with the Rankine-Hugoniot speed s given by

   $$s \;=\; \frac{f(\rho_l) - f(\rho_r)}{\rho_l - \rho_r} = \frac{\rho_l v_l - \rho_r v_r}{\rho_l - \rho_r} \tag{A.5}$$

   The coefficient v in (A.1a) will be *discontinuous* due to relation (A.3). We have $v_l > v_r$ in this case.

2. $\rho_l > \rho_r$
   The entropic solution is a rarefaction wave. Thus the coefficient v in (A.1a) will be smooth due to relation (A.3).

To obtain a simplified model and to apply Front–Tracking a rarefaction wave is approximated by a sequence of entropy violating shocks traveling with the Rankine–Hugoniot speed s, cf. (A.5). Figures A.1 and A.2 give a graphical representation of solutions to the Riemann problem (A.4). We depict for a given datum ρ_l possible states ρ_r that can be connected via a shock or a rarefaction.

The approximation of the rarefaction fan by a sequence of shock waves leads to a mathematically more interesting question. Due to relation (A.3) we obtain $v_l < v_r$ and therefore the OSLC is not satisfied. Then uniqueness of the Filippov characteristics is not guaranteed in general. Consequently, we will take a closer look into the solution structure in the rarefaction case. This will facilitate the discussion of reversible and duality solutions. However, we begin our investigation by considering the simpler case 1 in which the entropic solution to the LWR equation is a shock.

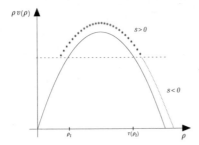

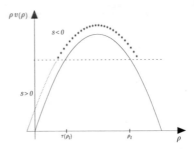

Figure A.1: Location of ρ_l and $\tau(\rho_l)$ and corresponding sign of shock speeds for entropic shock solutions.

Figure A.2: Location of ρ_l and $\tau(\rho_l)$ and corresponding sign of shock speeds for the (approximate) rarefaction case.

Shock solutions

We continue the investigation with a little more care, since in subsection 2.1.3 the shock-speed s was (often implicitly) assumed to be 0. Now we will allow $s \in \mathbb{R}$. We will see that only certain values of s play a role in our setup. This study will be helpful in the rarefaction case.

Let $\rho_l < \rho_r$ be fixed. The entropy solution [77] to this Riemann problem is a wave traveling with the Rankine-Hugoniot speed s (A.5) given by:

$$ s = \frac{f(\rho_l) - f(\rho_r)}{\rho_l - \rho_r} $$

This means that the coefficient v in the commodity equation changes with time, the discontinuity travels also with speed s. More explicitly, we have

$$ v(t, x) = \begin{cases} v_l & x - st < x_0 \\ v_r & x - st > x_0 \end{cases} \tag{A.6} $$

Now we can distinguish four cases depending on the relation of s, v_l and v_r as depicted in figures A.3 – A.6.

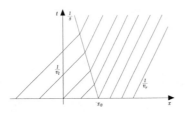

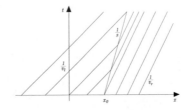

Figure A.3: Case 1

Figure A.4: Case 2

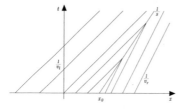

Figure A.5: Case 3

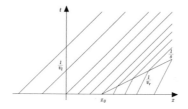

Figure A.6: Case 4

Note that since $v_l > v_r > 0$, we have $\frac{1}{v_l} < \frac{1}{v_r}$. The behavior of a solution is very different in the four cases. The first two correspond to transport along the uniquely defined characteristics as it was discussed in subsection 2.1.3 for $s = 0$. In the third case the inital data is also transported, but now it seems the mass is concentrating along the line of discontinuity of the coefficient v, i.e., along the line with slope $\frac{1}{s}$. We will discuss this situation in some more detail below. The last case corresponds to the opposite situation found in case one or two. In the latter cases the information is accelerated across the jump in v, in the fourth case it is decelerated.

However, we are going to show that in our situation only cases one and two can occur. This is done by a simple discussion of the possible relations of s given by (A.5), v_l and v_r. Note that for shocks we always have $v_l > v_r > 0$.

- Case 1: $\frac{1}{s} < 0 < \frac{1}{v_l} < \frac{1}{v_r}$
  The first case, i.e., a negative shock speed, is possible: Let $s < 0$.

$$\frac{\rho_l - \rho_r}{\rho_l \cdot v_l - \rho_r \cdot v_r} < \frac{1}{v_l} \quad \Leftrightarrow \quad \rho_l v_l - \rho_r v_l < \rho_l \cdot v_l - \rho_r \cdot v_r$$
$$\Leftrightarrow \quad 0 < \rho_r \cdot (v_l - v_r)$$

and similarly

$$\frac{\rho_l - \rho_r}{\rho_l \cdot v_l - \rho_r \cdot v_r} < \frac{1}{v_r} \quad \Leftrightarrow \quad \rho_l v_r - \rho_r v_r < \rho_l \cdot v_l - \rho_r \cdot v_r$$
$$\Leftrightarrow \quad 0 < \rho_l \cdot (v_l - v_r)$$

Hence, we require $\rho_r \neq 0$ and $\rho_l \neq 0$. Then $v_l > v_r$ ensures that both inequalities hold.

- Case 2: $0 < \frac{1}{v_l} < \frac{1}{v_r} < \frac{1}{s}$
  In case two we get by a similar computation as in (A.8) and (A.9) the inequalities

$$\rho_r(v_r - v_l) < 0 \qquad \rho_l(v_r - v_l) < 0 \qquad (A.7)$$

Obviously, if $\rho_r \neq 0$ and $\rho_l \neq 0$, (A.7) can be satisfied if $v_r < v_l$. If $\rho_l = 0$, we obtain by (A.5) $s = v_r$ and thus the reverse inequaltiy $v_r > s$, which is the starting point for the computations, does not hold. Thus condition (A.7) does not apply in this case.

124

- Case 3: $0 < \frac{1}{v_l} < \frac{1}{s} < \frac{1}{v_r}$

Here we also have $s > 0$, but the inequalities we obtain are different. We get

$$\frac{1}{s} > \frac{1}{v_l} \Leftrightarrow v_l > s \quad \Leftrightarrow \quad v_l > \frac{\rho_l \cdot v_l - \rho_r \cdot v_r}{\rho_l - \rho_r}$$
$$\Leftrightarrow \quad v_l \rho_l - v_l \rho_r < \rho_l v_l - \rho_r v_r$$
$$\Leftrightarrow \quad \rho_r(v_r - v_l) < 0 \qquad \text{(A.8)}$$

Again, we can assume $\rho_r > 0$ and with $v_r < v_l$ we satisfy the inequality. But we have another restriction on the shock speed:

$$\frac{1}{s} < \frac{1}{v_r} \Leftrightarrow v_r < s \quad \Leftrightarrow \quad v_r < \frac{\rho_l \cdot v_l - \rho_r \cdot v_r}{\rho_l - \rho_r}$$
$$\Leftrightarrow \quad v_r \rho_l - v_r \rho_r > \rho_l v_l - \rho_r v_r$$
$$\Leftrightarrow \quad \rho_l(v_r - v_l) > 0 \qquad \text{(A.9)}$$

In the case $\rho_l = 0$ we immediately get a contradiction. If $\rho_l > 0$, then we require $v_r > v_l$ for (A.9) to hold, leading to $\rho_r < \rho_l$ which is not possible for an entropic shock solution.

- Case 4: $0 < \frac{1}{s} < \frac{1}{v_l} < \frac{1}{v_r}$

We have $s > 0$ and consequently $\rho_l \cdot v_l - \rho_r \cdot v_r < 0$ (use (A.5) and $\rho_l < \rho_r$). Furthermore, we have

$$\frac{1}{s} < \frac{1}{v_l} \Leftrightarrow v_l < s \quad \Leftrightarrow \quad v_l < \frac{\rho_l \cdot v_l - \rho_r \cdot v_r}{\rho_l - \rho_r}$$
$$\Leftrightarrow \quad v_l \rho_l - v_l \rho_r > \rho_l v_l - \rho_r v_r$$
$$\Leftrightarrow \quad \rho_r(v_r - v_l) > 0 \qquad \text{(A.10)}$$

If $\rho_r = 0$, then $\rho_l < 0$ due to our general assumption, which clearly constitutes a contradiction to the intention in the design of the model. Hence we can assume $\rho_r > 0$ and we can conclude that for (A.10) to hold we require $v_r > v_l$. Due to the correspondence (A.3) in the LWR-model, this is equivalent to $\rho_r < \rho_l$ which presents a contradiction.

Remark A.1. *One can explicitly determine the regions/situations, for which cases one and two apply. Figure A.1 captures all the information. We require $\rho_r > \rho_l$ and $s < 0$ or $s > 0$, respectively. Since the shock speed is related to the fundamental diagram given by f as indicated by (A.5), we get:*

$$s < 0 \text{ requires } f(\rho_l) > f(\rho_r) \quad \Leftrightarrow \quad \rho_l < \sigma \text{ and } \rho_r > \tau(\rho_l)$$
$$s > 0 \text{ requires } f(\rho_l) < f(\rho_r) \quad \Leftrightarrow \quad \rho_l < \rho_r < \sigma \text{ or } \rho_l < \sigma, \ \sigma \leq \rho_r \leq \tau(\rho_l)$$

As the discussion shows only cases one and two can occur. Therefore the unique weak solution to the commodity equation is obtain by transporting the initial value along the uniquely defined Filippov characteristics.

Rarefaction solutions

Now we turn our attention to the more interesting case where $\rho_l > \rho_r$ and the entropy solution is a rarefaction wave. One idea to simplify the multi–commodity PDE–model (2.43) is to approximate a rarefaction wave by a (sequence of) shock wave(s) traveling with the Rankine-Hugionot speed given by (A.5). Since $\rho_l > \rho_r$, we have $v_l < v_r$ for the LWR-model due to relation (A.3) Note that the coefficient v in the commodity equation then does not satisfy the OSLC. Thus the uniqueness of the Filippov characteristics is not guaranteed a priori. As pointed out in subsection 2.1.3 we can in general not construct even a measure solution in this situation.

We will conduct a similar analysis as in the shock–case. As a result we solely have two possible scenarios for the characteristics. These are depicted in figures A.7 and A.8.

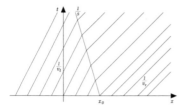

| Figure A.7: Case 1 | Figure A.8: Case 2 |

In the spirit of the entropic shock case (i.e., the one in which v satisfies the OSCL), we drew the characteristics with slope $\frac{1}{v_r}$ in the cone spanned by $y = \frac{1}{s}(x - x_0)$ and $y = \frac{1}{v_r}(x - x_0)$. We note that this procedure is motivated by the notion of reversible solutions discussed in [14]. We illustrate this method by an example.

We consider two examples presented in [93]. We want to compute a measure solution for the conservative equation

$$\partial_t \mu + \partial_x (v \, \mu) \quad = \quad 0 \tag{A.11a}$$
$$\mu(0, x) \quad = \quad \mu_0(x) \tag{A.11b}$$

In the first example, we choose the coefficient v as

$$v(t, x) \quad = \quad -sign(x)$$

This choice satisfies the OSLC and therefore the Filippov characteristics are unique. For the second example already mentioned in subsection 2.1.3 we choose

$$v(t, x) \quad = \quad +sign(x)$$

As we have already observed the Filippov characteristics are not unique in this setting. Figures A.9 and A.10 illustrate the different situations.

For $v = -sign(x)$ the unique measure solution $\mu(t, [a, b])$ can be obtained by tracing back the characteristics that lead through (t, a) and (t, b). In this fashion we obtain $(0, a^*)$ and $(0, b^*)$. The measure solution is then given by $\mu(t, [a, b]) = \mu_0([a^*, b^*])$.

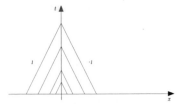
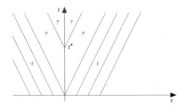

Figure A.9: Example for unique Filippov–characteristics for $v = -sign(x)$. However, solely a measure solution can be constructed by tracing back the characteristics.

Figure A.10: Example for non–unique Filippov–characteristics for $v = sign(x)$. Initial values can not simply be transported along the characteristics emanating at $\bar{x} = 0$. Conservation of mass can not be guaranteed by this procedure.

As pointed out in subsection 2.1.3 for $v = sign(x)$ the Filippov characteristics are not unique. The problem is that v does not satisfy the OSCL. However, if we rotate figure (A.10) and consider equation (A.11a) backwards in time, we are in the situation of figure (A.9). The coefficient v then satisfies the OSCL again and we can determine unique Filippov characteristics and therefore unique measure solutions p for a backward problem. These solutions p are called *reversible* solutions. These facilitate the definition of a solution to the forward problem (A.11a). This so–called duality solution can be characterized in terms of the unique Filippov characteristics for the backward problem, cf. Theorem 4.2.8 in [14]. If the initial measure posesses enough regularity (e.g., $\mu_0 \in L^1_{loc}(\mathbb{R})$ as outlined in subsection 2.1.3) we obtain the unique duality solution $\mu(t, [a, b])$ to the forward problem by following the Filippov characteristics for the backward problem. As in subsection 2.1.3 we obtain an interval $[a^*, b^*]$ at $t = 0$ such that $\mu(t, [a, b]) = \mu_0([a^*, b^*])$. More details and a thorough mathematical description of these concepts can be found in [14].

We investigate the procedure for the situation in figure A.7. Initially we obtain the forward characteristics depicted in figure A.11.

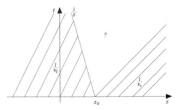
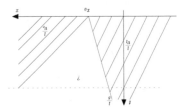

Figure A.11: Situation without backward Filippov characteristics. Since the OSLC is violated the generalized characteristics do not need to be unique.

Figure A.12: Construction of backward Filippov characteristics. This situation resembles the shock case discussed previously.

Note that in figure A.11 there is a region marked with a ? in which we do not know

127

how the generalized characteristics look. However, if we turn the figure as indicated in figure A.12 and solve (A.11a) backwards in time our coefficient v satisfies the OSCL again. The roles of v_r and v_l change. We have in the rarefaction case $\frac{1}{v_r} < \frac{1}{v_l}$. The backward problem therefore exaclty corresponds to the setup in figure A.3 (note that $\frac{1}{v_r}$ in figure A.12 correpsonds to $\frac{1}{v_l}$ in figure A.3). Consequently, we recover case 1 from the shock solution in which we know how the generalized characteristics are computed. This enables us to draw the characteristics given in figure A.7. The forward solution is then given by transport along the generalized characteristics which we obtain by considering figure A.12.

We conclude this appendix with a discussion of the feasible relations between velocities and shock–speed in the rarefaction case. This procedure is analogous to the one presented in the shock–case. Recall that for a rarefaction $v_r > v_l$ and approximate shock speed s is still given by (A.5)

- Case 1: $\frac{1}{s} < 0 < \frac{1}{v_r} < \frac{1}{v_l}$

  Since s is given by (A.5) and now $\rho_l > \rho_r$, we have

  $$s < 0 \quad \Leftrightarrow \quad f(\rho_l) - f(\rho_r) < 0$$

  This inequality can be satisfied. For a graphical proof we look at figure A.2. If $\sigma < \rho_l$ and $\tau(\rho_l) < \rho_r < \rho_l$, then we will have a shock travelling backwards. Of course, algebraic manipulations similar to the ones presented in the discussion of the shock cases lead to two inequalities which hold for the parameter settings in this case.

- Case 2: $0 < \frac{1}{v_r} < \frac{1}{v_l} < \frac{1}{s}$

  Again, s is given by (A.5) and $f(\rho) = \rho \cdot v$. We have

  $$\frac{1}{v_l} < \frac{1}{s} \quad \Leftrightarrow \quad s < v_l \quad \Leftrightarrow \quad \rho_l v_l - \rho_r v_r < rho_l v_l - \rho_r v_l$$
  $$\Leftrightarrow \quad 0 < \rho_r \cdot (v_r - v_l)$$
  $$\overset{\rho_r \geq 0}{\Leftrightarrow} \quad 0 < v_r - v_l \tag{A.12}$$

  (A.12) holds in our situation. If $\rho_r = 0$, then the shock speed s is given by $s = v_l$ (cf. (A.5)) and the case does not apply. Analogously one obtains

  $$\frac{1}{v_r} < \frac{1}{s} \quad \Leftrightarrow \quad s < v_r \quad \Leftrightarrow \quad \rho_l v_l - \rho_r v_r < rho_l v_r - \rho_r v_r$$
  $$\Leftrightarrow \quad 0 < \rho_l \cdot (v_r - v_l)$$
  $$\overset{\rho_l \geq 0}{\Leftrightarrow} \quad 0 < v_r - v_l \tag{A.13}$$

  Thus, (A.13) holds in this case. $\rho_l = 0$ implies (by $\rho_l > \rho_r$) $\rho_r < 0$, which does not make sense for the model.

- Case 3: $0 < \frac{1}{v_r} < \frac{1}{s} < \frac{1}{v_l}$

  The only thing different from (A.12) is the inequality sign. We obtain

  $$0 \quad > \quad v_r - v_l$$

  Thus, this case is irrelevant for our considerations as $v_r > v_l$.

- Case 4: $0 < \frac{1}{s} < \frac{1}{v_r} < \frac{1}{v_l}$

This case does not occur, either. The reasoning is similar to case 3.

Therefore, in our application the Filippov characteristics are unique even for $v_r > v_l$. This is due to the LWR–correspondence (A.3) and the positivity of the speeds which makes a situation as in figure A.10 impossible. Again, the unique weak solution to the multicommodity equation is obtained by transporting the initial value along the generalized characteristics.

Appendix B

Description of a Preprocessing Routine For the Multicommodity Model

As pointed out in subsection 2.1.2 we have a procedure to determine for a given network $G = (V, A)$ which values of the controls $\alpha_{\varepsilon\nu}^{i}(v;t), v \in V, \varepsilon \in \delta_v^-, \nu \in \delta_v^+$ are predetermined. Our method requires that we know *all* paths from a source $s \in \mathcal{S}$ to a destination $d \in \mathcal{D}$. Sven Krumke gave us an outline of on an algorithm which computes for a given pair $(s, d) \in \mathcal{S} \times \mathcal{D}$ *all* paths from s to d. The computation time of the procedure can be exponential in the number of nodes and is therefore not efficient in general. However, for the networks we consider it works well. The algorithm is briefly described here since it is vital for the performance and correctness of our simulation and for future work on optimization procedures.

We briefly recall some definitions from graph theory which can be found in [76], for example.

Definition B.1. (Adjacency matrix of a graph G): *A matrix A is called adjacency matrix for a graph $G = (V, E)$, if $A = \left(a_{ij}\right)_{i,j=1,\ldots,|V|}$, where*

$$a_{ij} \quad := \quad |\{e \in A \mid e = (v_i, v_j)\}|$$

Definition B.2. (Toplogical order): *Let $G = (V, A)$ be a directed graph. A toplogical order of G is a bijective mapping $\sigma : V \mapsto \{1, \ldots, |V|\}$ such that*

$$\sigma(u) \quad < \quad \sigma(v) \qquad \text{for all } (u, v) \in A$$

An implementation in pseudocode for a topological ordering of a graph $G = (V, A)$ can be found in [76], for example. It requires the graph to be given by its adjacency matrix or by adjacency lists. Furthermore, one needs to compute the inner degrees, i.e., the size of δ_v^-, for each node $v \in V$. The latter can be determined by algorithm 2.2 found in [76].

Example B.3. *Consider the simple network from figure B.1. Then the adjacency matrix*

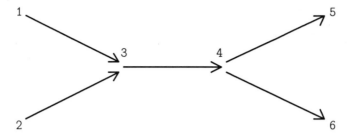

Figure B.1: A testnetwork with one junction of the First and one junction of the Second type.

A is given by

$$A = \begin{pmatrix} 0 & 0 & 1 & 0 & 0 & 0 \\ 0 & 0 & 1 & 0 & 0 & 0 \\ 0 & 0 & 0 & 1 & 0 & 0 \\ 0 & 0 & 0 & 0 & 1 & 1 \\ 0 & 0 & 0 & 0 & 0 & 0 \\ 0 & 0 & 0 & 0 & 0 & 0 \end{pmatrix}$$

A particular topological order is given by $\sigma(v) = v$ since we numbered the nodes accordingly already.

In order to determine shortest paths in a graph we need to associate a capacity function $c : A \to \mathbb{R}$ with the arcs of the network. Then we can measure the distance from a node $v \in V$ to a node $w \in V$. Here we can make use of the so–called Bellman equations, cf. for example pg. 186 in [76].

Definition B.4. (Bellman equations): *For a given capacity function $c : A \to \mathbb{R}$, the Bellman equations read:*

$$\begin{aligned} dist_c(s, s) &= 0 \\ dist_c(s, v) &= \min\{dist_c(s, u) + c(u, v), (u, v) \in A\} \text{ for } v \neq s \end{aligned}$$

Together with a topological order, they provide an easy formula to obtain the length of *all* paths from a given node $s \in V$ to a node $d \in V$. In particular, the shortest path can be determined.

We require that our graph does not contain loops for the following procedure to work. We describe it for a particular soure–destination pair $(s, d) \in \mathcal{S} \times \mathcal{D}$.

1. Sort the graph $G = (V, A)$ topologically.
   This is possible, since our graph does not contain loops. The time required is $\mathcal{O}(n^2)$ (if we use adjacency matrices; for a graph represented by its adjacency list the complexity is $\mathcal{O}(n + m)$). Herein, $n = |V|$ and $m = |A|$, i.e., n denotes the number of vertices in the graph and m the number of edges.

2. Determine for all vertices $v \in V$ the distance (shortest path) from v to d.
   With the aid of Bellman equations and the toplogical order this is possible in
   $\mathcal{O}(n^2)$. The capacity function we use is particularly simple. We choose $c \equiv 1$.
   At present we use the Dijkstra-algorithm to determine the shortest path.

3. Now enumerate all paths from s to d using ENUMERATEPATH(s,d) defined by
   ENUMERATEPATH(x,y)
   for all successors $v_1, \ldots, v_p$ of x
   if there is a path from v_j to y

$$E_i = x + ENUMERATEPATH(v_j, y)$$

   end % if
   end % for

The cost of the algorithm should be $\mathcal{O}(|P|)$ for every path P with $|P|$ arcs, which is optimal asymptotically.

The output of ENUMERATEPATH(s,d) are **all** paths leading from s to d. Therefore, if we run the routine for all $(s, d) \in \mathcal{S} \times \mathcal{D}$, i.e., for all commodities $i = 1, \ldots, I = |\mathcal{S} \times \mathcal{D}|$, we know the sets $\mathcal{P}_i, i = 1, \ldots, I$ from subsection 2.1.2. Then we can use the procedure outlined in subsection 2.1.2 to determine the values of some controls $\alpha^i_{\varepsilon\nu}(v; t)$.

Appendix C

Computation of the Reduced Hessian for the Instantaneous Control Problem

In this appendix we formally apply a framework presented in [102] to compute the Hessian for the instantaneous control problem from subsection 2.2.2. Once the Hessian is known we can apply Newton–like methods in the optimization process which should lead to superlinear or even locally quadratic convergence. As we will see below, the computation of the Hessian is expensive. In particular we need to solve an additional partial differential equation. At present it is not known whether the overall run–time of our algorithm will improve by adding second–order information.

C.1 A general procedure

The algorithm presented below is taken from [102]. The notation is adapted to our situation. Additionally, we provide an explanation for the proposed formulas.

We recall the framework discussed in subsection 2.2.1. We are concerned with a problem of the form

$$\min J(y, u) \tag{C.1a}$$
$$\text{subject to}$$
$$c(y, u) = 0 \tag{C.1b}$$
$$u \in U_{ad} \tag{C.1c}$$

J is the so–called objective functional, $y \in Y$ is the state–variable, $u \in U_{ad} \subset U$ is an applicable control with a feasible control set U_{ad} and c is the so–called state–operator. Equation (C.1b) is known as *equation of state*.

In the sequel we will assume that the following holds:

A1 $U_{ad} \subset U$ is convex, bounded, closed and nonempty.

A2 For all $u \in U_{ad}$ the state equation $c(y, u) = 0$ has a unique solution $y = y(u)$, which is continuously Fréchet–differentiable.

A3 $c_y(y, u) \in \mathcal{L}(Y, Z)$ has a bounded inverse for all
$(y, u) \in W_{ad} := \{(y, u) \in Y \times U \mid u \in U_{ad}, c(y, u) = 0\}$

A4 $J : Y \times U \mapsto \mathbb{R}$ and $c : Y \times U \mapsto Z$ are twice continuously Fréchet differentiable.

In [102] we find an algorithm for the computation of the reduced Hessian in direction $v \in U_{ad}$. We set $z := \partial_u y(u)v$ for $u, v \in U_{ad}$. The function L appearing below is the Lagrangian function (2.113). The proposed scheme reads:

1. Compute the state $y = y(u) \in Y$ by solving the state equation

$$c(y, u) \;=\; 0$$

2. Compute the adjoint state $p = p(u) \in Z^*$ by solving the adjoint equation

$$c_y^*(y(u), u)\, p \;=\; -J_y(y(u), u)$$

3. Compute $z = z(u) \in Y$ as solution of the linearized state equation

$$c_y(y(u), u)\, z \;=\; -c_u(y(u), u)v$$

4. Compute $h = h(u) \in Z^*$ by solving the adjoint system

$$c_y^*(y(u), u)\, h \;=\; -\partial_{uy}^2 L(y(u), u, p(u))\, v - \partial_{yy}^2 L(y(u), u, p(u))\, z$$

5. Set

$$j''(u)v \;=\; \partial_{uu}^2 L(y(u), u, p(u))\, v + \partial_{yu}^2 L(y(u), u, p(u))\, z + c_u^*(y(u), u)\, h(u)$$

The goal of this section is to show where the formulas come from and why they are reasonable. In the next section we will use them to derive an expression for the Hessian of the instantaneous control problem. We need to recall some of the results from subsection 2.2.1.

We were able to establish first order necessary optimality conditions for problem (C.1), cf. Corollary 2.2.9:

Corollary C.1. *Let $(\bar{y}, \bar{u})$ be an optimal solution to (C.1) and let assumptions A1 to A3 hold. Then there exists an adjoint state $\bar{p} \in Z^*$ such that the following optimality conditions hold*

$$
\begin{aligned}
c(\bar{y}, \bar{u}) &= 0 & \bar{u} \in U_{ad} & \qquad \text{(C.2a)} \\
c_y(\bar{y}, \bar{u})^* \bar{p} &= -J_y(\bar{y}, \bar{u}) & & \qquad \text{(C.2b)} \\
\langle J_u(\bar{y}, \bar{u}) + c_u(\bar{y}, \bar{u})^* \bar{p}, u - \bar{u} \rangle_{U^*, U} &\geq 0 & \forall u \in U_{ad} & \qquad \text{(C.2c)}
\end{aligned}
$$

Sufficient conditions involve second order information stemming from the Hessian or an approximation thereof. As for the gradient expression, we consider the reduced objective functional $\hat{J}(u)$, cf. (2.110), which is related to the Lagrangian (2.113) via

$$\hat{J}(u) \;=\; J(y(u), u) = L(y(u), u, p) = J(y(u), u) + \langle p, c(y(u), u)\rangle_{Z^*, Z} \qquad \forall p \in Z^*$$

Therefore, we formally have

$$\hat{J}'(u) \;=\; D_u L(y(u), u, p), \qquad p \in Z^* \tag{C.3a}$$
$$\hat{J}''(u) \;=\; D_{uu} L(y(u), u, p) = D_u(D_u L(y(u), u, p)), \qquad p \in Z^* \tag{C.3b}$$

Note that by D_u we denote the total derivative w.r.t. u. We have seen in section 2.2.1 that relation (C.3a) was the essential ingredient to obtain the gradient equation. Similarly, equation (C.3b) enables us to compute the Hessian. However, we need to clarify our notation to some degree.

The derivative $\hat{J}'(u)$ is a linear operator, since the reduced cost fuctional $\hat{J}$ is a map from U to $\mathbb{R}$. Therefore, we formally have for $u, v \in U_{ad} \subset U$ and $r \in U$

$$\mathbb{R} \ni \hat{J}'(u+v)r - \hat{J}'(u)r \;\;=\;\; (J''(u)v)r + \mathcal{O}(\|v\|^2) \tag{C.4a}$$
$$\overset{(C.3b)}{=} \;\; (D_{uu} L(y(u), u, p)v)r + \mathcal{O}(\|v\|^2) \tag{C.4b}$$

In this fashion we can compute the directional derivative of $\hat{J}'(u)$ in direction $r \in U$. For a general $p \in Z^*$ we found that we have (cf. (2.115))

$$\hat{J}'(u) \;=\; D_u L(y(u), u, p) = (\partial_u y(u))^* \partial_y L(y(u), u, p) + \partial_u L(y(u), u, p) \tag{C.5}$$

A suitable choice of $p \in Z^*$ ensures that $\partial_y L(y(u), u, p) = 0$ and we can then compute the gradient.

For a general $p \in Z^*$ we have by (C.5)

$$\hat{J}'(u+v)r + \hat{J}'(u)r \;=\; (\partial_u y(u+v))^* \partial_y L(y(u+v), u+v, p)r - (\partial_u y(u))^* \partial_y L(y(u), u, p)r$$
$$+ \partial_u L(y(u+v), u+v, p)r - \partial_u L(y(u), u, p)r$$

If $L(y, u, p)$ is twicely Fréchet–differentiable w.r.t. $y \in Y$ and $u \in U_{ad}$, we formally obtain

$$\hat{J}'(u+v)r + \hat{J}'(u)r \;=\; (\partial_u y(u))^* \partial_{yy}^2 L(y(u), u, p) \partial_u y(u) v r \tag{C.6a}$$
$$+ (\partial_u y(u))^* \partial_{uy}^2 L(y(u), u, p) v r \tag{C.6b}$$
$$+ \partial_{yu}^2 L(y, u, p) \partial_u y(u) v r + \partial_{uu}^2 L(y, u, p) v r + \mathcal{O}(\|v\|^2) \tag{C.6c}$$

Remark C.2. *One can make these considerations more rigorous. In particular one needs to ensure that the operators $\partial_y L(y, u, p)$ and $\partial_u L(y, u, p)$ are twicely continuously Fréchet–differentiable and that the second–order operators and their application to the occuring elements are well defined. This is a lengthy process and we just give one example for these considerations here:*

In the first component, L maps some $y \in Y$ to $\mathbb{R}$. Therefore, we have

$$\partial_y L(\cdot, u, p) : Y \mapsto \mathcal{L}(Y, \mathbb{R}) = Y^* \qquad Y \ni w \to \partial_y L(w, u, p) \in \mathcal{L}(Y, \mathbb{R})$$

137

In words, $\partial_y L(w, u, p)$ is a linear and bounded map from Y to $\mathbb{R}$ for every $w \in Y$. We could also write this matter more clearly as duality product for some $v \in Y$

$$\partial_y L(w, u, p)v \;=\; \langle v, \partial_y L(w, u, p)\rangle_{Y,Y^*}.$$

Similarly, we have $\partial^2_{yy} L(w, u, p) \in \mathcal{L}(Y, \mathcal{L}(Y, \mathbb{R}))$ for every $w \in Y$. Therefore, we have for some $\mu, \nu \in Y$

$$\mathbb{R} \ni \partial^2_{yy} L(w, u, p)\mu\,\nu \;=\; \langle \nu, \partial^2_{yy} L(w, u, p)\mu \rangle_{Y,Y^*}.$$

Additionally, we have for the control–to–state map $U \ni u \mapsto y(u) \in Y$

$$y_u(\cdot) : U \mapsto \mathcal{L}(U, Y) \qquad \tilde{u} \mapsto y_u(\tilde{u})$$

Therefore, we have for $v \in U$ that $y_u(\tilde{u})v \in Y$ for all $\tilde{u} \in U$. Furthermore, we have $(y_u(\tilde{u}))^* \in \mathcal{L}(Y^*, U^*)$. Consequently, we obtain a more explicit and clearer notation

$$
\begin{aligned}
(\partial_u y(u))^* \partial^2_{yy} L(y(u), u, p)\partial_u y(u)vr \;&=\; \langle r, (\partial_u y(u))^* \partial^2_{yy} L(y(u), u, p)\partial_u y(u)v\rangle_{U,U^*} \\
&=\; \langle \partial_u y(u)r, \partial^2_{yy} L(y(u), u, p)(\partial_u y(u)v)\rangle_{Y,Y^*}.
\end{aligned}
$$

In order to obtain the formulas from the algorithm stated above, we need to find an expression for the operator

$$(\partial_u y(u))^* : Y^* \to U^* \qquad Y^* \ni q \mapsto (\partial_u y(u))^* q \in U^*$$

One particularly attractive option is to linearize the state equation (C.2a) in direction $v \in U_{ad}$. We formally obtain up to first order

$$c_y(y(u), u)y_u(u)v + c_u(y(u), u)v \;=\; 0 \in Z \qquad\qquad \text{(C.7)}$$

Remark C.3. *For a linear and bounded operator $A : X \mapsto Y$ the dual or adjoint operator A^*, which we assume exists, has the signature $A^* : Y^* \mapsto X^*$, since by definition we have for $y^* \in Y^*$ and $x \in X$*

$$y^*(Ax) =: \langle Ax, y\rangle_{Y,Y^*} \;=\; \langle x, A^* y^*\rangle_{X,X^*}.$$

This reasoning also shows that we can deduce from $c_u(y(u), u) \in \mathcal{L}(U, Z)$ that $c_u^*(y(u), u) \in \mathcal{L}(Z^*, U^*)$.

We derive from (C.7) for some $\lambda \in Z^*$

$$
\begin{aligned}
\mathbb{R} \ni 0 \;&=\; \langle c_y(y(u), u)y_u(u)v + c_u(y(u), u)v, \lambda\rangle_{Z,Z^*} \\
&=\; \langle v, \big(y_u^*(u)c_y^*(y(u), u) + c_u^*(y(u), u)\big)\,\lambda\rangle_{U,U^*}.
\end{aligned}
$$

and therefore we formally have

$$\partial_u y^*(u) \;=\; -c_u^*(y(u), u)(c_y^*(y(u), u))^{-1} \in Y^* \qquad\qquad \text{(C.8)}$$

We finally have all the ingredients to justify the formulas in the presented algorithm. We define for a direction $v \in U_{ad}$

$$z = z(u) := \partial_u y(u)v \quad \in Y$$

Steps 3 to 5 in the algorithm are just a clever way to compute formula (C.6). By (C.7) we have

$$c_y(y(u), u)z = -c_u(y(u), u)v$$

which is step 3. Step 4 and step 5 combine the application of $\partial_u y^*(u)$ and the summation in (C.6). However, they are exececuted in two steps. Step 4 partially evaluates the expressions (C.6a) and (C.6b), as formula (C.8) involves the inversion of an operator which is not practial explicitly. Therefore, we introduce another variable $h = h(u)$ as the solution of

$$(c_y^*(y(u), u))h = -\partial_{yy}^2 L(y(u), u, p)z - \partial_{uy}^2 L(y(u), u, p)v$$

Then we obtain finally

$$\hat{J}''(u)v = c_u^*(y(u), u)h + \partial_{yu}^2 L(y, u, p)z + \partial_{uu}^2 L(y, u, p)v$$

C.2 Application to the Instantaneous Control Problem

Here we apply the rather abstract formulas from the previous section to the instantaneous control problem from subsection 2.2.2. The expressions for the reduced Hessian at a single junction can be used to compute the Hessian on the whole network. We assume that the network $G = (V, A)$ consists of D dispersing junction $d \in \{1, \ldots, D\}$ where the ingoing road is labeled 1_d and the two outgoing roads are labeled 2_d and 3_d, respectively. We set $\vec{\alpha} = (\alpha^1, \ldots, \alpha^D)$ and assume the applicable control at a dispersing junction d is α^d. Let the cost–functional in the optimization be of the form

$$J(\vec{rho}, \vec{\alpha}) = \frac{\gamma}{4} \tau \sum_{d=1}^{D} (\alpha^d - \bar{\alpha}^d)^2 + \sum_{j \in A} \int_{a_j}^{b_j} \frac{\tau}{2} \mathcal{F}_j(\rho_j(x)) \, dx$$

Furthermore, assume that z and h satisfy equations (C.22), (C.23) and (C.24), (C.25) at a dispersing junction $d \in \{1, \ldots, D\}$. μ denotes the solution of the adjoint equations (C.18) at this junction. Then the reduced Hessian in direction $\vec{v} = (v_1, \ldots, v_d)$ reads on a time interval $I := [t, t + \Delta t]$

$$\hat{J}''(\vec{\alpha})\vec{v} = \left(\hat{J}''(\alpha^1)v_1, \ldots, \hat{J}''(\alpha^D)v_D \right)$$

where the expression $\hat{J}''(\alpha^d)v_d$ at a dispersing junction d is given by (cf. (C.26))

$$\begin{aligned} \hat{J}''(\alpha^d)v_d = {} & \frac{\gamma}{2} \tau v_d \, \mathrm{id}_{1b} + f_1'(\rho_{1_d}(b_{1_d})) \cdot z_{1_d}(b_{1_d}) \, \mathrm{id}_{1b} \left(\mu_{3_d}(a_{3_d}) - \mu_{2_d}(a_{2_d}) \right) \\ & + f_{1_d}(\rho_{1_d}(b_{1_d})) \cdot \left(h_{3_d}(a_{3_d}) - h_{2_d}(a_{2_d}) \right) \mathrm{id}_{1b} \end{aligned}$$

C.2.1 Dispersing Junction

First we discuss the case of a dispersing junction, i.e., we have one ingoing road 1 and two outgoing roads labeled 2 and 3, respectively. We consider a time interval $I_i := [t_i, t_{i+1}]$

139

and set $\rho^i_j(x) = \rho_j(x, t_i)$. On a time interval I_i, the optimal control problem (2.144) can be stated as

$$\min_{\alpha^{i+1}} J_{i+1}(\vec{\rho}^{\,i+1}, \alpha^{i+1}) \text{ subject to} \tag{C.9a}$$

$$\frac{\rho^{i+1}_j(x) - \rho^i_j(x)}{\tau_{i+1}} + \frac{\partial}{\partial x} f_j(\rho^{i+1}_j(x)) = 0, \quad j \in \{1, 2, 3\} \tag{C.9b}$$

$$f_2(\rho^{i+1}_2(a_2)) - \alpha^{i+1} f_1(\rho^{i+1}_1(b_1)) = 0 \tag{C.9c}$$

$$f_3(\rho^{i+1}_3(a_3)) - (1 - \alpha^{i+1}) \cdot f_1(\rho^{i+1}_1(b_1)) = 0 \tag{C.9d}$$

$$\rho^i_j(x) = g_j(x) \tag{C.9e}$$

$$f_1(\rho^{i+1}_1(a_1)) = h_{i+1} \tag{C.9f}$$

$$0 \le \alpha^{i+1} \le 1 \tag{C.9g}$$

where the objective functional is given by

$$J_{i+1}(\vec{\rho}^{\,i+1}, \alpha^{i+1}) := \frac{\gamma}{4} \left(\tau_{i+1}(\alpha^{i+1} - \bar{\alpha})^2 + \tau_{i+1}(\alpha^i - \bar{\alpha})^2 \right) \tag{C.10a}$$

$$+ \sum_{j=1}^{3} \int_{a_j}^{b_j} \frac{\tau_{i+1}}{2} \left(\mathcal{F}_j(\rho^{i+1}_j(x)) + \mathcal{F}_j(\rho^i_j(x)) \right) dx \tag{C.10b}$$

In the sequel we will drop the time indices for ease of notation, setting $(\rho_1, \rho_2, \rho_3) = \vec{\rho} := \vec{\rho}^{\,i+1}$, $\vec{g}(x) := \vec{\rho}^{\,i}(x)$ and $\alpha := \alpha^{i+1}$. Note that the term (C.10a) is used for steering the system to a predifened state with corresponding control $\bar{\alpha}$. If one is not interested in this kind of control, one can set $\gamma = 0$ and one obtains a "classical" optimal control problem.

In the notation from the previous section $y = \vec{\rho}^{\,i+1}$, $u = \alpha$. Furthermore, $U_{ad} = \{\alpha \in U \mid 0 \le \alpha \le 1\}$. For the space U we can simply choose $U = \mathbb{R}$ in our problem; however, below it will prove helpful to consider $U = L^p$. This is not a restriction here, since our control $\alpha \in U_{ad}$ is constant and for any bounded interval I we therefore have $\alpha \in L^p(I)$. Since we have a hyperbolic system (or more accurately, a time discretized version of one), we can at most expect the solution to be of bounded variation or belong to some L^p-space (shocks are in principal possible).

We define $\tau := \tau_{i+1}$. Then the operator $c(y, u)$ takes the form:

$$c(\vec{\rho}, \alpha) = \begin{pmatrix} \frac{\rho_1(x) - g_1(x)}{\tau} + \frac{\partial}{\partial x} f_1(\rho_1(x)) \\ \frac{\rho_2(x) - g_2(x)}{\tau} + \frac{\partial}{\partial x} f_2(\rho_2(x)) \\ \frac{\rho_3(x) - g_3(x)}{\tau} + \frac{\partial}{\partial x} f_3(\rho_3(x)) \\ f_2(\rho_2(a_2)) - \alpha \cdot f_1(\rho_1(b_1)) \\ f_3(\rho_3(a_3)) - (1 - \alpha) \cdot f_1(\rho_1(b_1)) \\ f_1(\rho_1(a_1)) - h \end{pmatrix} \tag{C.11}$$

where the functions $g_j(x), j \in 1, 2, 3$ are given; in particular, they do not depend on the control α. $h = h_{i+1} \in \mathbb{R}$ is an intial value for the problem on road 1, cf. (C.9f). Without it, the system (C.9) does not possess a unique solution.

Remark C.4. *In the following we assume $\rho_j(x), \alpha \in L^p$. For α this is not a restriction. However, for ρ_j this assumption is not sufficient since its derivative $\partial_x \rho_j$ occurs which needs to be suitably defined. We can not guarantee that $\rho_j(x)$ belongs to some Sobolev–space either since entropic shock–waves can occur in the solution. Therefore, the results in this section are mathematically not rigorous in the way they are stated.*

The choice of the space L^p for some p is due to the fact that its dual space $(L^p)^*$ is readily determined: $(L^p)^* = L^{p'}$ where $\frac{1}{p} + \frac{1}{p'} = 1$, cf. [1, 106]. We will use this fact to point out to which spaces some of the occuring operators belong.

Bearing remark C.4 in mind, the operator c in (C.11) has the following "signature":

$$c : (L^p)^3 \times L^p \;\mapsto\; \left[(L^p)^6\right]^* = \left(L^{p'}\right)^6 \tag{C.12}$$

where p' is the dual exponent for p.

Now it is clear, that the Lagrangian multiplier w from the previous section needs to belong to $(L^p)^6$. In the sequel we will set $w = \mu$.

Remark C.5. *Again, note that we have computed the adjoint equations for the problem (C.9) in section 2.2.2. These equations contain terms involving $\partial_x \mu_j = \partial_x w_j$, thus the regularity requirement $w_j = \mu_j \in L^p$ is not sufficient for a rigorous mathematical treatment.*

For a given α, we can of course obtain a solution $\vec{\rho}$ to equations (C.9b) - (C.9f). Therefore, we have $\vec{\rho} = \vec{\rho}(\alpha)$. Our reduced objective function then reads:

$$\hat{J}(\alpha) \;=\; J(\vec{\rho}(\alpha), \alpha)$$

and the Lagrangian becomes

$$L(\vec{\rho}, \alpha, \mu) \;=\; J(\vec{\rho}, \alpha) + \langle c(\vec{\rho}, \alpha), \mu \rangle_{(L^{p'})^6, (L^p)^6}$$

$$= \; J(\vec{\rho}, \alpha) + \sum_{i=1}^{6} \int_{a_j}^{b_j} c_j(\vec{\rho}, \alpha) \mu_j(x) \, dx$$

With these identifications one can compute all the operators occuring in section C.1. This is a rather lengthy process, so we give an example of the procedure for one of the terms below. The remaining ones will just be stated.

We explicitly compute $\partial_\alpha L(\vec{\rho}, \alpha, \mu)$. Due to the linear structure of the Lagrangian we simply need to find expression for the quantities

$$\partial_\alpha J(\vec{\rho}, \alpha) \in \mathcal{L}(U, \mathbb{R}) \qquad \partial_\alpha c(\vec{\rho}, \alpha) \in \mathcal{L}(U, (L^{p'})^6)$$

We obtain

$$\partial_\alpha J(\vec{\rho}, \alpha) \;=\; \frac{\gamma}{2} \cdot \tau(\alpha - \bar{\alpha}) \cdot \mathrm{id}_1 \tag{C.13a}$$

$$\partial_\alpha c(\vec{\rho}, \alpha) \;=\; (0, 0, 0, -f_1(\rho_1(b_1)) \cdot \mathrm{id}_2, -f_1(\rho_1(b_1)) \cdot \mathrm{id}_2, 0)^T \tag{C.13b}$$

where we assume the maps $\mathrm{id}_1 : L^p \mapsto L^p$ and $\mathrm{id}_2 : L^p \mapsto L^{p'}$ to be suitably defined.

Remark C.6. *In particular we note that the control $\alpha \in [0,1] \subset \mathbb{R}$ and the we assume for the control–space $U = \mathbb{R} \subset L^p(I)$ for some bounded interval I. For a non–constant function $\alpha \in L^p$ the choice of the objective functional (C.10) is unsuitable, since we require $H_{i+1}(\vec{\rho}, \alpha) \in \mathbb{R}$. A better choice in this case would be*

$$J_{i+1}(\vec{\rho}, \alpha^{i+1}) \quad := \quad \frac{\gamma}{2} \int_{t_i}^{t_{i+1}} \left((\alpha^{i+1} - \bar{\alpha})^2\right) \, dt \tag{C.14a}$$

$$+ \sum_{j=1}^{3} \int_{a_j}^{b_j} \frac{\tau_{i+1}}{2} \left(\mathcal{F}_j(\rho_j^{i+1}(x)) + \mathcal{F}_j(\rho_j^{i}(x))\right) dx \tag{C.14b}$$

Consequently, we would obtain.

$$\partial_{\alpha^{i+1}} J_{i+1}(\vec{\rho}, \alpha^{i+1}) \quad = \quad \gamma \int_{t_i}^{t_{i+1}} (\alpha^{i+1} - \bar{\alpha}) \cdot \mathrm{id}_1 \, dt \quad \in \mathcal{L}(L^p, \mathbb{R}) \tag{C.15}$$

Summarizing, the partial derivative of the Lagrangian w.r.t. α reads

$$\mathcal{L}(L^p, \mathbb{R}) \ni \partial_{\alpha} L(\vec{\rho}, \alpha, \mu) \quad = \quad \frac{\gamma}{2} \cdot \tau(\alpha - \bar{\alpha}) \cdot \mathrm{id}_1 \tag{C.16a}$$

$$+ f_1(\rho_1(b_1))(\langle \mathrm{id}_2, \mu_5 \rangle_{L^{p'}, L^p} - \langle \mathrm{id}_2, \mu_4 \rangle_{L^{p'}, L^p} \tag{C.16b}$$

Remark C.2.1. *The multiplier $w = \mu$ is determined by the adjoint equation. Its derivation in this framework is more involved. We will give the formulas below.*

It can be shown that we have

$$\partial_{\vec{\rho}} J(\vec{\rho}, \alpha) \quad = \quad \begin{pmatrix} \frac{\tau}{2} \mathcal{F}'_1(\rho_1(x)) \\ \frac{\tau}{2} \mathcal{F}'_2(\rho_2(x)) \\ \frac{\tau}{2} \mathcal{F}'_3(\rho_3(x)) \end{pmatrix} \quad \in \left(L^{p'}\right)^3 \tag{C.17}$$

and additionally for $(\mu \in L^p)^6$

$$\left(\partial_{\vec{\rho}} c(\vec{\rho}, \alpha)\right)^* \mu \quad = \quad \begin{pmatrix} \langle \square_1, \frac{\mu_1}{\tau} - f'_1(\rho_1(x)) \cdot \frac{\partial}{\partial x} \mu_1 \rangle + f'_1(\rho_1(x)) \cdot \square_1(x) \cdot \mu_1(x) \big|_{a_1}^{b_1} \\ \langle \square_2, \frac{\mu_2}{\tau} - f'_2(\rho_2(x)) \cdot \frac{\partial}{\partial x} \mu_2 \rangle + f'_2(\rho_2(x)) \cdot \square_2(x) \cdot \mu_2(x) \big|_{a_2}^{b_2} \\ \langle \square_3, \frac{\mu_3}{\tau} - f'_3(\rho_3(x)) \cdot \frac{\partial}{\partial x} \mu_3 \rangle + f'_3(\rho_3(x)) \cdot \square_3(x) \cdot \mu_3(x) \big|_{a_3}^{b_3} \\ (f'_2(\rho_2(a_2)) \cdot \square_2(a_2) - \alpha \cdot f'_1(\rho_1(b_1)) \cdot \square_1(b_1))) \cdot \langle 1, \mu_4 \rangle \\ (f'_3(\rho_3(a_3)) \cdot \square_3(a_3) - (1 - \alpha) \cdot f'_1(\rho_1(b_1)) \cdot \square_1(b_1)) \cdot \langle 1, \mu_5 \rangle \\ f'_1(\rho_1(a_1)) \cdot \square_1(a_1) \langle 1, \mu_6 \rangle \end{pmatrix}$$

We recall that $\partial_{\vec{\rho}} c(\vec{\rho}, \alpha) \in \mathcal{L}((L^p)^3, (L^{p'})^6)$, therefore $(\partial_{\vec{\rho}} c(\vec{\rho}, \alpha))^* \in \mathcal{L}((L^p)^6, (L^{p'})^3)$ and consequently $(\partial_{\vec{\rho}} c(\vec{\rho}, \alpha))^* w \in (L^{p'})^3$. The symbols $\square_z, z = 1, 2, 3$ then denote the zth componenct of an element $(q_1, q_2, q_3) = q \in (L^p)^3$ in the (well–defined) application of the operator $(\partial_{\vec{\rho}} c(\vec{\rho}, \alpha))^*$ to the element q.

Plugging these expressions into (C.2b) leads to the adjoint system (cf. (2.146))

$$\frac{\tau}{2} \mathcal{F}'_j(\rho_j(x)) + \frac{\mu_j}{\tau} - f'_j(\rho_j(x)) \cdot \frac{\partial}{\partial x} \mu_j \quad = \quad 0 \quad j = 1, 2, 3 \tag{C.18a}$$

$$\mu_1(b_1) = \alpha \mu_2(a_2) + (1 - \alpha) \mu_3(a_3) \quad = \quad \alpha \langle 1, \mu_4 \rangle + (1 - \alpha) \langle 1, \mu_5 \rangle \tag{C.18b}$$

$$\mu_3(b_3) = 0 \qquad \mu_2(b_2) = 0 \tag{C.18c}$$

$$\mu_1(a_1) \quad = \quad \langle 1, \mu_6 \rangle \tag{C.18d}$$

Remark C.7. *Equation (C.18d) seems somewhat strange, we have not derived such a condition before. We have to note that the number $\langle 1, \mu_6 \rangle$ is* **not** *specified, i.e., the value of $\mu_1(a_1)$ is not fixed. The adjoint system is backwards in space, so we use (C.18a) for $j = 2, 3$ and (C.18c). This determines $\mu_1(b_1)$ due to (C.18b). If $\mu_1(a_1)$ would be fixed, we would have two boundary values on road 1 and the ODE for road 1 might not have a solution. Since we have the freedom how to choose $\langle 1, \mu_6 \rangle$, we proceed as follows. Solve (C.18a) for $j = 1$ with boundary condition given by (C.18b). Then we obtain $\mu_1(a_1)$. Then in the correct multiplier $\mu = \mu(\alpha)$, we set for its last component $\mu_6 = \frac{1}{b_1 - a_1} \cdot \mu_1(a_1) \in \mathbb{R}$.*

For the Hessian we need the terms $\partial^2_{\alpha\alpha} L$, $\partial^2_{\alpha\bar\rho} L = \partial_\alpha \partial_{\bar\rho} L$, $\partial^2_{\bar\rho\alpha} L = \partial_{\bar\rho} \partial_\alpha L$ and $\partial^2_{\bar\rho\bar\rho} L$. Again, we are more explicit for the easiest example. We have by (C.16)

$$
\mathcal{L}(L^p, \mathbb{R}) \ni \partial_\alpha L(\vec\rho, \alpha, \mu) = \frac{\gamma}{2} \cdot \tau(\alpha - \bar\alpha) \cdot \mathrm{id}_1
$$
$$
+ f_1(\rho_1(b_1))(\langle \mathrm{id}_2, \mu_5 \rangle_{L^{p'}, L^p} - \langle \mathrm{id}_2, \mu_4 \rangle_{L^{p'}, L^p}
$$

Consequently, $\partial_\alpha \partial_\alpha L(\vec\rho, \alpha, \mu) = \partial^2_{\alpha\alpha} L(\vec\rho, \alpha, \mu) \in \mathcal{L}(L^p, \mathcal{L}(L^p, \mathbb{R}))$ and therefore we need to apply two arguments $v, r \in U \subset L^p$ to obtain a real number:

$$
\partial^2_{\alpha\alpha} L(\vec\rho, \alpha, \mu) v \in \mathcal{L}(L^p, \mathbb{R}) \quad \text{and} \quad \left(\partial^2_{\alpha\alpha} L(\vec\rho, \alpha, \mu) v \right) r \in \mathbb{R}
$$

We have for a direction $v \in U \subset L^p$

$$
\partial_\alpha L(\vec\rho, \alpha + v, \mu) r - \partial_\alpha L(\vec\rho, \alpha, \mu) r = \frac{\gamma}{2} \cdot \tau \cdot v \cdot r
$$

and therefore we obtain

$$
\partial^2_{\alpha\alpha} L(\vec\rho, \alpha, \mu) = \frac{\gamma}{2} \tau \cdot \mathrm{id}_{1a} \cdot \mathrm{id}_{1b}
$$

where $\mathrm{id}_{1a} : L^p \mapsto L^p$ acts on the direction $v \in L^p$ and $\mathrm{id}_{1b} : L^p \mapsto L^p$ on the argument $r \in L^p$.

For $\vec\kappa, \vec\lambda \in L^{p3}$ and $r \in L^p$ a similar reasoning leads to

$$
\left([\partial_{\bar\rho} \partial_\alpha L(\vec\rho, \alpha, \mu)] \vec\kappa \right) r = f_1'(\rho_1(b_1)) \cdot \kappa_1(b_1) \cdot r \cdot (\mu_3(a_3) - \mu_2(a_2))
$$
$$
\left([\partial_\alpha \partial_{\bar\rho} L(\vec\rho, \alpha, \mu)] r \right) \vec\kappa = r \cdot f_1'(\rho_1(b_1)) \cdot \kappa_1(b_1) \cdot (\mu_3(a_3) - \mu_2(a_2))
$$
$$
\left([\partial_{\bar\rho} \partial_{\bar\rho} L(\vec\rho, \alpha, \mu)] \vec\lambda \right) \vec\kappa = \left\langle \left(\begin{array}{c} \frac{\tau}{2} \mathcal{F}''_1(\rho_1(x)) \cdot \lambda_1(x) - f_1''(\rho_1(x)) \cdot \lambda_1(x) \cdot \frac{\partial}{\partial x} \mu_1 \\ \frac{\tau}{2} \mathcal{F}''_2(\rho_2(x)) \cdot \lambda_2(x) - f_2''(\rho_2(x)) \cdot \lambda_2(x) \cdot \frac{\partial}{\partial x} \mu_2 \\ \frac{\tau}{2} \mathcal{F}''_3(\rho_3(x)) \cdot \lambda_3(x) - f_3''(\rho_3(x)) \cdot \lambda_3(x) \cdot \frac{\partial}{\partial x} \mu_3 \end{array} \right), \left(\begin{array}{c} \kappa_1 \\ \kappa_2 \\ \kappa_3 \end{array} \right) \right\rangle
$$

We turn our attention to step 3 of the algorithm from section C.1. We define

$$
z_j(x; v) := \frac{\partial}{\partial \alpha} \rho_j(\alpha, x) v \tag{C.21}
$$

It can be shown formally that the equation $c_y(y(u), u) z = -c_u(y(u), u) v$ in our case reads for $j = 1, 2, 3$

$$
\frac{1}{\tau} \cdot z_j(x; v) + f_j'(\rho_j(\alpha, x)) \cdot \frac{\partial}{\partial x} z_j(x; v) + f_j''(\rho_j(\alpha, x)) \cdot \frac{\partial}{\partial x} \rho_j(\alpha, x) \cdot z_j(x; v) = 0 \tag{C.22}
$$

The corresponding boundary conditions are given by

$$z_1(a_1; v) = 0 \tag{C.23a}$$

$$f_2'(\rho_2(\alpha, a_2)) \cdot z_2(a_2; v) = f_1(\rho_1(\alpha, b_1)) \cdot v + \alpha f_1'(\rho_1(\alpha, b_1)) \cdot z_1(b_1; v) \tag{C.23b}$$

$$f_3'(\rho_3(\alpha, a_3)) \cdot z_3(a_3; v) = -f_1(\rho_1(\alpha, b_1)) \cdot v + (1 - \alpha) f_1'(\rho_1(\alpha, b_1)) \cdot z_1(b_1; v) \tag{C.23c}$$

The adjoint equation with a modified right hand side from step 4 of the algorithm from section C.1 reads in our case

$$\frac{h_j}{\tau} - f_j'(\rho_j(x)) \cdot \frac{\partial}{\partial x} h_j = \left(f_j''(\rho_j(x)) \frac{\partial}{\partial x} \mu_j - \frac{\tau}{2} F_j''(\rho_j(x)) \right) z_j(x) \quad j = 1, 2, 3 \tag{C.24}$$

Note that μ_j, the solution to the adjoint equations (C.18), enters the equations explicitly. Furthermore, we need the solution of equations (C.22). The boundary conditions for equations (C.24) are given by

$$h_1(b_1) = (\alpha + v)h_2(a_2) + ((1 - \alpha) - v)h_3(a_3) \tag{C.25a}$$

$$h_3(b_3) = 0 \qquad h_2(b_2) = 0 \tag{C.25b}$$

$$h_1(a_1) = \langle 1, h_6 \rangle \tag{C.25c}$$

Note that the direction $v \in L^p$ enters explicitly in the coupling conditions. Additionally, we remark that we find

$$\langle 1, h_4 \rangle = h_2(a_2) \quad \text{and} \quad \langle 1, h_5 \rangle = h_3(a_3)$$

We can solve equations (C.24) for h_j, $j = 1, 2, 3$ starting on roads 2 and 3. The coupling conditions then give us the initial value $h_1(b_1)$ on road 1. This enables us to solve the equation for $j = 1$.

In the final step of the algorithm from section C.1 we need to compute the following expression which in our case reduces to

$$c_\alpha^*(\vec{\rho}(\alpha), \alpha) h(\alpha) = f_1(\rho_1(b_1)) \cdot (h_3(a_3) - h_2(a_2)) \operatorname{id}_{1b}$$

The reduced Hessian in direction $v \in L^p$ is then given by

$$\hat{J}''(\alpha)v = \frac{\gamma}{2} \tau v \operatorname{id}_{1b} + f_1'(\rho_1(b_1)) \cdot z_1(b_1) \operatorname{id}_{1b} (\mu_3(a_3) - \mu_2(a_2)) \tag{C.26a}$$

$$+ f_1(\rho_1(b_1)) \cdot (h_3(a_3) - h_2(a_2)) \operatorname{id}_{1b} \tag{C.26b}$$

where $\vec{z} = (z_1, z_2, z_3)$ is the solution to (C.22), (C.23) and $\vec{h} = (h_1, h_2, h_3)$ is the solution to (C.24), (C.25).

C.2.2   Merging Junction

We give the results for a merging junction, i.e., we have two ingoing roads 1 and 2 and one outgoing road labeled 3, respectively. Not surprisingly, the formulas and equations from the previous subsection will not change all that much. The only modification occurs in the coupling conditions. This case is investigated to be able to extend the computation of the reduced Hessian to a network.

We consider a time interval $I_i := [t_i, t_{i+1}]$ and set $\rho_j^i(x) = \rho_j(x, t_i)$. On a time interval I_i, the optimal control problem (2.144) can be stated as

$$\min_{\alpha^{i+1}} J_{i+1}(\vec{\rho}^{\,i+1}, \alpha^{i+1}) \text{ subject to} \tag{C.27a}$$

$$\frac{\rho_j^{i+1}(x) - \rho_j^i(x)}{\tau_{i+1}} + \frac{\partial}{\partial x} f_j(\rho_j^{i+1}(x)) = 0, \quad j \in \{1, 2, 3\} \tag{C.27b}$$

$$f_1(\rho_1^{i+1}(b_1)) + f_2(\rho_2^{i+1}(b_2)) = f_3(\rho_3^{i+1}(a_3)) \tag{C.27c}$$

$$\rho_j^i(x) = g_j(x) \tag{C.27d}$$

$$f_1(\rho_1^{i+1}(a_1)) = h_{i+1} \tag{C.27e}$$

$$f_1(\rho_1^{i+1}(a_1)) = h_{i+1} \tag{C.27f}$$

where the objective functional is given by

$$J_{i+1}(\vec{\rho}^{\,i+1}, \alpha^{i+1}) := \frac{\gamma}{4} \left(\tau_{i+1}(\alpha^{i+1} - \bar{\alpha})^2 + \tau_{i+1}(\alpha^i - \bar{\alpha})^2 \right) \tag{C.28a}$$

$$+ \sum_{j=1}^{3} \int_{a_j}^{b_j} \frac{\tau_{i+1}}{2} \left(\mathcal{F}_j(\rho_j^{i+1}(x)) + \mathcal{F}_j(\rho_j^i(x)) \right) dx \tag{C.28b}$$

We skip the detailed analysis as it is similar to the dispersing case. We obtain for the adjoint equations:

$$\frac{\tau}{2} \mathcal{F}'_j(\rho_j(x)) + \frac{\mu_j}{\tau} - f'_j(\rho_j(x)) \cdot \frac{\partial}{\partial x} \mu_j = 0 \quad j = 1, 2, 3 \tag{C.29a}$$

$$\mu_3(b_3) = 0 \tag{C.29b}$$

$$\mu_2(b_2) = \mu_3(a_3) \tag{C.29c}$$

$$\mu_1(b_1) = \mu_3(a_3) \tag{C.29d}$$

For the gradient of the reduced objective functional we obtain

$$\hat{J}'(\alpha) = \frac{\gamma}{2} \tau(\alpha - \bar{\alpha}) \, \mathrm{id}_1$$

Step 3 in the algorithm reads in the merging case with $z_j = z_j(x; v) := \frac{\partial}{\partial \alpha} \rho_j(\alpha, x) v$

$$\frac{1}{\tau} \cdot z_j(x; v) + f'_j(\rho_j(\alpha, x)) \cdot \frac{\partial}{\partial x} z_j(x; v) + f''_j(\rho_j(\alpha, x)) \cdot \frac{\partial}{\partial x} \rho_j(\alpha, x) \cdot z_j(x; v) = 0 \tag{C.30}$$

The corresponding boundary conditions are given by

$$z_1(a_1; v) = 0 \tag{C.31a}$$

$$z_2(a_2; v) = 0 \tag{C.31b}$$

$$f'_3(\rho_3(\alpha, a_3)) \cdot z_3(a_3; v) = f'_1(\rho_1(\alpha, b_1)) \cdot z_1(b_1; v) + f'_2(\rho_2(\alpha, b_2)) \cdot z_2(b_2; v) \tag{C.31c}$$

In step 4 we have to determine h as solution of

$$\frac{h_j}{\tau} - f'_j(\rho_j(x)) \cdot \frac{\partial}{\partial x} h_j = \left(f''_j(\rho_j(x)) \frac{\partial}{\partial x} \mu_j - \frac{\tau}{2} F''_j(\rho_j(x)) \right) z_j(x) \quad j = 1, 2, 3 \tag{C.32}$$

Note that μ_j, the solution to the adjoint equations (C.29), enters the equations explicitly. Furthermore, we need the solution of equations (C.31). The boundary conditions for equations (C.32) are given by

$$h_3(b_3) = 0 \qquad\qquad\qquad (\text{C.33a})$$
$$h_2(b_2) = h_3(a_3) \qquad\qquad (\text{C.33b})$$
$$h_1(b_1) = h_3(a_3) \qquad\qquad (\text{C.33c})$$

Finally we obtain for the reduced Hessian in direction $v \in U \subset L^p$

$$\hat{J}''(\alpha)v = \frac{\gamma}{2}\tau v \; \mathrm{id}_{1b}$$

This result is not surprising as we have no "real" control at this node. In particular we note that for $v \in (0,1] \subset U_{ad} \subset \mathbb{R}$ this expression is always positive.

Remark C.8. *Note that $c_\alpha(\rho,\alpha) \equiv 0$ here and therefore the term $(c_\alpha(\rho,\alpha))^*h$ does not contribute to the expression in the Hessian. Consequently, we* **do not need** *to solve equations (C.32) for a merging junction to compute the reduced Hessian (which is pointless anyway; if the network consisted of a single merging junction, the question for an optimal control can not be posed, since there is no control variable in our model). In a network, we need the information on h on every road since it propagates backwards through the network. Therefore, the solution of (C.32) is in general* **necessary**!

Bibliography

[1] H. W. Alt, *Lineare Funktionalanalysis*, Springer-Verlag, Berlin, Heidelberg, New York, 1999.

[2] D. Armbruster, C. de Beer, M. Freitag, T. Jagalski, and C. Ringhofer, *Autonomous control of production networks using a pheromone approach*. to appear in PHYSICA, 2006.

[3] D. Armbruster, P. Degond, and C. Ringhofer, *Kinetic and fluid models for supply chains supporting policy attributes*. submitted to Transp. Theory and Stat. Phys., 2004.

[4] ——, *A model for the dynamics of large queuing networks and supply chains*, SIAM J. Applied Mathematics, 48 (2006), pp. 896–920.

[5] D. Armbruster, D. Marthaler, and C. Ringhofer, *Kinetic and fluid model hierarchies for supply chains*, SIAM J. on Multiscale Modeling, 2 (2004), pp. 43–61.

[6] A. Aw, A. Klar, M. Materne, and M. Rascle, *Derivation of continuum flow traffic models from microscopic follow the leader models*, SIAM J. Appl. Math., 63 (2002), pp. 259–289.

[7] A. Aw and M. Rascle, *Resurrection of second order models of traffic flow?*, SIAM J. Appl. Math., 60 (2000), pp. 916–944.

[8] P. Bagnerini and M. Rascle, *A multi-class homogenized hyperbolic model of traffic flow*, SIAM J. Math. Anal., 35 (2003), pp. 949–973.

[9] C. Bardos, A. LeRoux, and J. Nedelec, *First order quasilinear equations with boundary conditions*, Comm. in Partial Diff. Eqns., 4 (1979), p. 1017.

[10] S. Benzoni-Gavage and R. M. Colombo, *An n-population model for traffic flow*, European Journal of Applied Mathematics, 63 (2003), pp. 818–833.

[11] M. Berggren, *Numerical solution of a flow-control problem: vorticity reduction by dynamic boundary action*, SIAM J. Sci. Comput., 19 (1998), p. 829.

[12] P. Bogacki and L. F. Shampine, *A 3(2) pair of runge-kutta formulas*, Appl. Math. Letters, 2 (1989), pp. 1–9.

[13] J. F. Bonnans, J. C. Gilbert, C. Lemarechal, and C. A. Sagastizabal, *Numerical Optimization*, Springer, Berlin, Heidelberg, New York, 1997.

[14] F. BOUCHUT AND F. JAMES, *One-dimensional transport equations with discontinuous coefficients*, Nonlinear Analysis, Theory, Methods and Applications, 32 (1998), p. 891.

[15] A. BRESSAN, *Hyperbolic Systems of Conservation Laws*, no. 20 in Oxford lecture series in mathematics and its applications, Oxford University Press, Oxford, New York, 2000.

[16] M. BURGER AND R. PINNAU, *Fast optimal design of semiconductor devices*, SIAM J. Appl. Math., 64 (2003), pp. 108–126.

[17] P. H. CALAMAI AND J. J. MORE, *Projected gradient methods for linearly constrained problems*, Math. Programming, 39 (1987), p. 93.

[18] H. CHOI, R. TEMAM, P. MOIN, AND J. KIM, *Feedback control for unsteady flow and its applications to the stochastic burgers equation*, Journal of Fluid Mechanics, 253 (1993), p. 509.

[19] G. COCLITE, M. GARAVELLO, AND B. PICCOLI, *Traffic flow on road networks*, SIAM J. Math. Anal., 36 (2005), pp. 1862–1886.

[20] R. M. COLOMBO, *A 2×2 hyperbolic traffic flow model*, Math. Comput. Modelling, 35 (2002), pp. 683–688.

[21] ——, *Hyperbolic Phase Transitions in Traffic Flow*, SIAM J. on Appl. Math., 63 (2002), pp. 708–721.

[22] R. R. D. GAZIS, R. HERMAN, *Nonlinear follow-the-leader model of traffic flow*, Oper. Res., 9 (1961), p. 545.

[23] C. M. DAFERMOS, *Polygonal approximations of solutions of the initial value problem for a conservation law*, J. Math. Anal. Appl., 38 (1972), p. 33.

[24] ——, *Hyperbolic conservation laws in continuum physics*, Springer, Berlin, Heidelberg, New York, 2000.

[25] C. F. DAGANZO, *Requiem for second order fluid approximation of traffic flow*, Transp. Research B, 29 (1995), p. 277.

[26] ——, *A behavioral theory of multi-lane traffic flow part i: Long homogeneous freeway sections*, Trans. Res. B, 36 (2002), pp. 131–158.

[27] ——, *A behavioral theory of multi-lane traffic flow part ii: Merges and the onset of congestion*, Trans. Res. B, 36 (2002), pp. 159–169.

[28] C. F. DAGANZO AND J. LAVAL, *Lane changing in traffic streams*, Trans. Res. B, 40 (2006), pp. 251–264.

[29] C. D'APICE AND R. MANZO, *A fluid dynamic model for supply chains*, Networks and Heterogenous Media (NHM), 1 (2006), pp. 379–398.

[30] I. C. DIVISION. Information available at URL http://www.cplex.com. 889 Alder Avenue, Suite 200, Incline Village, NV 89451, USA.

[31] A. F. FILIPPOV, *Differential equations with discontinuous right-hand side*, A.M.S. Transl., 42 (1964), p. 199.

[32] A. FÜGENSCHUH, S. GÖTTLICH, M. HERTY, A. KLAR, AND A. MARTIN, *A discrete optimization approach to large scale supply networks based on partial differential equations.* submitted, 2006.

[33] A. FÜGENSCHUH, M. HERTY, A. KLAR, AND A. MARTIN, *Combinatorial and continuous models for the optimization of traffic flow networks*, SIAM J. Optimization, 16 (2006), pp. 1155–1176.

[34] M. GARAVELLO AND B. PICCOLI, *Source-destination flow on a road network*, Comm. Math. Sci., 3 (2005), pp. 261–283.

[35] ——, *Traffic flow on a road network using the aw-rascle model*, Comm. Partial Differential Equations, 31 (2006), pp. 243–275.

[36] I. GASSER, G. SIRITO, AND B. WERNER, *Bifurcation analysis of a class of car–following traffic models*, Physica D, 197 (2004), pp. 222–241.

[37] E. GODLEWSKI AND P. RAVIART, *Numerical Approximation of Hyperbolix Systems of Conservation Laws*, no. 118 in Applied Mathematical Sciences, Springer, New York, Berlin, Heidelberg, 1996.

[38] S. GÖTTLICH, M. HERTY, AND A. KLAR, *Network models for supply chains*, Comm. Math. Sci., 3 (2005), pp. 545–559.

[39] ——, *Modelling and optimization of supply chains on complex networks*, Comm. Math. Sci., 4 (2006), pp. 315–330.

[40] J. GREENBERG, *Extension and amplification of the aw-rascle model*, SIAM J. Appl. Math., 62 (2001), pp. 729–745.

[41] J. GREENBERG, A. KLAR, AND M. RASCLE, *Congestion on multilane highways*, SIAM J. Appl. Math., 63 (2003), pp. 818–833.

[42] A. GRIEWANK AND G. F. CORLISS. Information available at URL http://www.autodiff.org.

[43] M. GUGAT, G. LEUGERING, K. SCHITTKOWSKI, AND E. J. P. G. SCHMIDT, *Combinatorial and continuous models for the optimization of traffic flow networks*, in Online optimization of large scale systems, M. Grötschel, S. O. Krumke, and J. Rambau, eds., Springer, 2006, pp. 251–270.

[44] M. GUGAT, G. LEUGERING, AND E. J. P. G. SCHMIDT, *Global controllability between steady supercritical flows in channel networks*, Math. Meth. Appl. Sci., 27 (2004), pp. 781–802.

[45] M. HEINKENSCHLOSS, *A time-domain decomposition iterative method for the solution of distributed linear quadratic optimal control problem*, Journal of Computational and Applied Mathematics, 173 (2005), p. 169.

[46] D. HELBING AND A. GREINER, *Modeling and simulation of multi–lane traffic flow*, Physical Review E, 55 (1997), p. 5498.

[47] D. HELBING AND M. SCHRECKENBERG, *Cellular automata simulating experimental properties of traffic flows*, Physical Review E, 59 (1999), pp. R2505–R2508.

[48] M. HERTY, *Mathematics of Traffic Flow Networks: Modeling, Simulation and Optimization*, PhD thesis, TU Darmstadt, 2004.

[49] M. HERTY, M. GUGAT, A. KLAR, AND G. LEUGERING, *Optimal control for traffic flow networks*, Journal of Optimization Theory and Applications (JOTA), 126 (2005), pp. 589–616.

[50] M. HERTY, R. ILLNER, A. KLAR, AND V. PANFEROV, *Qualitative properties of solutions to systems of fokker-planck equations for multilane traffic flow*. to appear in Transport Theory and Statistical Physics, 2006.

[51] M. HERTY AND A. KLAR, *Modelling, simulation and optimization of traffic networks*, SIAM J. Sci. Comp., 25 (2004), pp. 1066–1087.

[52] M. HERTY, A. KLAR, AND L. PARESCHI, *General kinetic models for vehicular traffic flow and Monte Carlo methods*, CMS, 5 (2005), pp. 155–169.

[53] M. HERTY, S. MOUTARI, AND M. RASCLE, *Optimization criteria for modeling intersections of vehicular traffic flow*, Networks and Heterogenous Media, 1 (2006), pp. 275–294.

[54] M. HERTY AND M. RASCLE, *Coupling conditions for a class of "second-order" models for traffic flow*, SIAM J. Math. Anal., 38 (2006), pp. 595–616.

[55] M. HINZE, *Optimal and instantenous control of the instantionary Navier-Stokes equations*. Habilitation Thesis, 2000. Fachbereich Mathematik, Technische Universität Berlin.

[56] M. HINZE AND A. KAUFFMANN, *On a distributed control law with an application to the control of unsteady flow around a cylinder*, International Series of Numerical Mathematics, 133 (1999), p. 177.

[57] M. HINZE AND K. KUNISCH, *Conrol strategies for fluid flows - optimal vs suboptimal control*, in ENUMATH 97, H. G. B. et. al., ed., 1997, p. 351.

[58] ——, *4 control strategies for the Navier–Stokes equations*, in ESIAM: Proceedings, vol. 4, 1998, p. 181.

[59] ——, *Three control methods for time-dependent fluid flow*, Flow, Turbulence and Combustion, 65 (2000), p. 273.

[60] ——, *Second order methods for optimal control of time-dependent fluid flow*, Siam J. Control and Optim., 40 (2001), pp. 925–946.

[61] M. HINZE AND R. PINNAU, *An Optimal Control Approach to Semiconductor Design*, Mathematical Models and Methods in Applied Sciences (M3AS), 12 (2002), pp. 89–107.

150

[62] H. Holden and N. H. Risebro, *A mathematical model of traffic flow on a network of unidirectional roads*, SIAM J. Math. Anal., 26 (1995), pp. 999–1017.

[63] ——, *Front Tracking for Hyperbolic Conservation Laws*, no. 152 in Applied Mathematical Sciences, Springer, New York, Berlin, Heidelberg, 2002.

[64] R. Hundhammer, *Simulation und Echtzeitsteuerung sschwingender Saitennetzwerke*, PhD thesis, Fachbereich Mathematik, Technische Universität Darmstadt, 2002.

[65] R. Illner, C. Kirchner, and R. Pinnau, *A derivation of the aw–rascle traffic models from fokker–planck type kinetic models.* submitted to Quarterly of Applied Mathematics, 2007.

[66] R. Illner, A. Klar, and M. Materne, *Vlasov-fokker-planck models for multilane traffic flow*, Comm. Math. Sci., 1 (2003), pp. 1–12.

[67] W. L. Jin and Z. Xin, *The relaxation schemes for systems of conservation laws in arbitrary space dimensions*, Comm. Pure Appl. Math., 48 (1995), pp. 235–255.

[68] F. John, *Partial Differential Equations*, vol. 1 of Applied Mathematical Sciences, Springer-Verlag, Berlin, Heidelberg, New York, 4 ed., 1982.

[69] C. T. Kelley, *Iterative methods for optimization*, no. 18 in Frontiers in Applied Mathematics, SIAM, 1999.

[70] B. S. Kerner, *Congested traffic flow*, Transp. Res. Rec., 1678 (1998), pp. 160–167.

[71] ——, *Experimental features of self–organization in traffic flow*, Phys. Rev. Lett., 81 (1998), pp. 3797–3800.

[72] A. Klar, R. Kühne, and R. Wegener, *Mathematical models for vehicular traffic*, Surv. Math. Ind., 6 (1996), p. 215.

[73] A. Klar and R. Wegener, *A hierarchy of models for multilane vehicular traffic i: Modeling*, SIAM J. Appl. Math., 59 (1998), p. 983.

[74] ——, *Kinetic derivation of macroscopic anticipation models for vehicular traffic*, SIAM J. Appl. Math., 60 (2000), pp. 1749–1766.

[75] M. H. A. Klar, *Simplified dynamics and optimization of large scale traffic networks*, Mathematical Modelling and Methods in Applied Science, 14 (2004), pp. 579–601.

[76] S. O. Krumke and H. Noltemeier, *Graphentheoretische Konzepte und Algorithmen*, Leitfäden der Informatik, Teubner Verlag, 2005.

[77] N. S. Kruzkov, *First order quasi linear equations in several independent variables*, Math. USSR Sbornik, 10 (1970), pp. 217–243.

[78] J. L. Lions, *Optimal Control of Systems Governed by Partial Differential Equations*, Springer–Verlag, Berlin, 1971.

[79] G. LAGNESE AND G. LEUGERING, *Time-domain decomposition of optimal control problems for the wave equation*, System Control Lett., 48 (2003), pp. 229–242.

[80] D. LAX, *Hyperbolic systems of conservation laws and the mathematical theory of shock waves*, CBMF-NSF Regional Series in Applied Math., SIAM, Philadelphia, 1973.

[81] J. P. LEBACQUE, *Les modéles macroscopiques du traffic*, Annales des Pont., 67 (1993), pp. 24–45.

[82] J. P. LEBACQUE AND M. KHOSHYARAN, *First order macroscopic traffic flow models: Intersection modeling, network modeling*, in Proceedings of the 16th International Symposium on Transportation and Traffic Theory (ISTTT), Elsevier, 2005, pp. 365–386.

[83] R. LeVEQUE, *Numerical Methods for Conservation Laws*, Lectures in mathematics: ETH Zürich, Birkhäuser-Verlag, Basel, 1992.

[84] C. D. LEVERMORE, *Moment closure hierarchies for kinetic theories*, J. Stat. Phys., 83 (1996), pp. 1021–1065.

[85] M. J. LIGHTHILL AND G. B. WHITHAM, *On kinematic waves*, Proc. Royal Soc. Edinburgh, 229 (1955), pp. 281–316.

[86] A. MAY, *Traffic Flow Fundamentals*, Prentice Hall, Englewood Cliffs, NJ, 1990.

[87] D. Q. MAYNE AND H. MICHALSKA, *Receding horizon control of nonlinear systems*, IEEE Transactions on Automatic Control, 35 (1990), pp. 814–824.

[88] J. J. MORÉ AND G. TORALDO, *Algorithms for bound constraint quadratic programming problems*, Num. Math., 55 (1989), p. 377.

[89] ——, *On the solution of large quadratic programming problems with bound constraints*, SIAM J. on Opt., 1 (1991), pp. 93–113.

[90] S. G. NASH AND A. SOFER, *Linear and nonlinear programming*, The McGraw–Hill Companies, New York, St. Louis, San Francisco, 1996.

[91] H. J. PAYNE, *Freflo: A macroscopic simulation model for freeway traffic*, Transportation Research Record, 722 (1979), pp. 68–77.

[92] R. PINNAU AND G. THÖMMES, *Optimal Boundary Control of Glass Cooling Processes*, Math. Methods Appl. Sci., 27 (2004), pp. 1261–1281.

[93] F. POUPAUD AND M. RASCLE, *Measure-solutions to the linear multi-dimensional transport equation with non-smooth coefficients*, Comm. Part. Diff. Eq., 22 (1997), p. 337.

[94] I. R. RAFATOV, D. D. SIJACIC, AND U. EBERT., *Spatio-temporal patterns in a semiconductor–gas–discharge system: stability analysis and full numerical solution*. submitted to Phys. Rev. E., 2006.

[95] P. I. RICHARDS, *Shock waves on the highway*, Operations Research, 4 (1956), pp. 42–51.

[96] R. ROTHERBY, *Car following models*, in Traffic Flow Theory – A State of the Art Report, Special Report 165, Washington, DC, 1975, Transportation Research Board. Chapter 4.

[97] P. SPELLUCCI, *Numerische Verfahren der nichtlinearen Optimierung*, Birkhäuser Verlag, Basel, Boston, Berlin, 1993.

[98] B. TEMPLE, *Systems of conservation laws with invariant submanifolds*, Trans. Amer. Math. Soc., 280 (1983), pp. 781–795.

[99] F. TRÖLTZSCH, *Optimale Steuerung Partieller Differentialgleichungen*, Vieweg, Wiesbaden, 2005.

[100] M. ULBRICH, *Nonsmooth Newton–like methods for variational inequalities and constrained optimization problems in function spaces*, habilitation thesis, Technische Universität München, Zentrum für Mathematik, 2001.

[101] ——, *Semismooth Newton methods for operator equations in function spaces*, SIAM J. Optim., 13 (2003), pp. 805–842.

[102] S. ULBRICH, *Optimal control of nonlinear hyperbolic conservation laws with source terms*, habilitation thesis, Technische Universität München, 2002.

[103] ——, *A sensitivity and adjoint calculus for discontinuous solutions of hyperbolic conservation laws with source terms*, SIAM Journal on Control and Optimization, 41 (2002), pp. 740–797.

[104] ——, *Adjoint-Based Derivative Computations for the Optimal Control of Discontinuous Solutions of Hyperbolic Conservation Laws.*, Systems & Control Letters, 48 (2003), pp. 309–324.

[105] G. C. K. WONG AND S. C. WONG, *A multi–class traffic flow model – an extension of the LWR model with heterogeneous drivers*, Transp. Research A, 36 (2002), pp. 827–841.

[106] K. YOSIDA, *Functional Analysis*, Springer–Verlag, Berlin, Göttingen, Heidelberg, 1965.

[107] M. ZHANG, C.-W. SHU, G. C. K. WONG, AND S. C. WONG, *A weighted essentially non-oscillatory numerical scheme for a multi-class lighthill-whitham-richards traffic flow model*, Journal of Computational Physics, 191 (2003), pp. 639–659.

[108] P. ZHANG, R. LIU, S. WONG, AND S. DAI, *Hyperbolicity and kinematic waves of a class of multi-population partial differential equations*, European Journal of Applied Mathematics, 60 (2005), p. 916.

Curriculum Vitae

Present address:

    Claus Kirchner, Im Dunkeltälchen 39, 67663 Kaiserslautern, Germany

Date and Place of Birth: May 29, 1979, Gießen, Germany.

Country of Citizenship: Federal Republic of Germany.

Education:

10/2004 – Date:	PhD Candidate, FB Mathematik, TU Kaiserslautern, Kaiserslautern
03/2006 – 10/2006:	PhD Research, University of Washington, Seattle, WA, USA
08/2004:	Diploma thesis in Mathematics with A. Klar
10/1999 – 08/2004:	Studies in Mathematics and Computer Science, TU Darmstadt, Darmstadt
11/1998 – 10/1999:	Military service
08/1998 – 10/1998:	Job at Rosconi Design + Funktion GmbH
1989 – 07/1999:	High school, Gymnasium Philippinum Weilburg, Weilburg
1986 – 1989:	Elementary school

Academic Awards:

10/2004 – Date:	Fellowship granted by the German Research Foundation (DFG)
04/2006 – 07/2006:	Fellowship granted by the German Academic Exchange Service (DAAD)
09/2004:	Best student research project in 2004 at KOM, TU Darmstadt, Darmstadt

Work Experience:

11/2006 – Date:	Research and Teaching Assistant, FB Mathematik, TU Kaiserslautern
10/2004 – 02/2006:	Research and Teaching Assistant, FB Mathematik, TU Kaiserslautern
10/2001 – 02/2004:	Research and Teaching Assistant, FB Mathematik, TU Darmstadt
10/2000 – 08/2000:	Internship, Volksbank Wetzlar–Weilburg eG, Wetzlar

Kaiserslautern, August 7, 2007   Claus Kirchner